DATE DUE

D0855605

I daylie see many that delight to looke on Mappes but yet for want of skill in Geography, they knowe not with what manner of lines they are traced, nor what those lines do signify nor yet the true use of Mappes.

Thomas Blundeville, A Brief Description of Universal Mappes and Cardes and Their Use *(London 1589)*

An Examination of Cartography in Relation to Culture and Civilization

MAPS & MAN

Norman J. W. Thrower
Professor of Geography
University of California
Los Angeles

Prentice-Hall, Inc., Englewood Cliffs, New Jersey

To Page, Anne, and Mary

ISBN: P 0-13-555953-7
C 0-13-555961-8

Library of Congress Catalog Card Number: 70–166141

10 9 8 7 6 5 4 3 2

Prentice-Hall International, Inc., *London*
Prentice-Hall of Australia Pty., Ltd., *Sydney*
Prentice-Hall of Canada Ltd., *Toronto*
Prentice-Hall of India Private Limited, *New Delhi*
Prentice-Hall of Japan, Inc., *Tokyo*

CONTENTS

PREFACE

This is a book about maps rather than about mapping. Although there are necessary references to map-making, these are included only so that the reader may be better informed about the maps which result from the use of particular methods. It is not in any sense a "how-to-do-it" work, of which a number of good examples already exist. The writer believes that those who wish to make maps must learn by actually creating them, and that this involves long hours of practice and years of professional experience.

However, one can discover a great deal about maps by studying them; hence, the illustrations in this work are most important. For a majority of people, even geographers, a knowledge of maps rather than the principles of map-making is needed. It is to provide information on the nature and development of maps and the lure and lore of cartography that this book has been prepared. As Edmond Halley, the English astronomer who was himself a distinguished cartographer observed in 1686, by the use of the map certain phenomena "...may be better understood, than by any verbal description whatsoever."

In presenting this study, the author would like to take the opportunity of thanking his professional colleagues located in many different centers for their help. Specific contributions are indicated in the text or in the footnotes but others, including some by former and current students, which are more difficult to acknowledge individually, are nevertheless appreciated. The University of California, and particularly its Los Angeles campus, has provided an academic climate which made the book possible. Resources of many institutions were utilized but, on a regular basis, especially those of the UCLA Libraries—Research, William Andrews Clark, Powell, Map, and Special Collections. Anna Lang, Peggy Burbank, Tricia Caldwell, and Noel Diaz contributed in different and special ways to the work. My wife Betty provided the encouragement needed to sustain me as I wrote these thoughts on the subject which has occupied most of my professional life.

Norman J. W. Thrower
Los Angeles, California

As a branch of human endeavor, cartography has a long and interesting history which well reflects the state of cultural activity, as well as man's perception of the world, in different periods. The maps of early man were attempts to depict earth distributions graphically in order to better visualize them; like those of primitive peoples, these maps served specific needs. Viewed in its development through time, the map is a sensitive indicator of the changing thought of man, and few of his works seem to be such an excellent mirror of culture and civilization. In the modern world the map performs a number of significant functions, among which are its use as: a necessary tool in the comprehension of spatial phenomena; a most efficient device for the storage of information, including three-dimensional data; and a fundamental research tool permitting an understanding of distributions and relationships not otherwise known or imperfectly understood. A knowledge of maps and their contents is not automatic—it has to be learned; and it is important for educated people to know about maps even though they may not be called upon to make them. The map is one of a select group of communications media without which, McLuhan has suggested, "the world of modern science and technologies would hardly exist."[1]

Cartography, like architecture, has attributes of both a scientific and an artistic pursuit, a dichotomy which is certainly not satisfactorily reconciled in all

[1] H. Marshall McLuhan, *Understanding Media* (New York: McGraw-Hill, 1964), esp. pp. 157–58. This book, and other writings by this author, have much relevance to cartography as a means of communication, even though we may not accept the full implications of the phrase, "the medium is the message."

Introduction

1

presentations. Some maps are successful in their display of material but are scientifically barren, while in others an important message may be obscured because of the poverty of representation. An amazing variety of maps exist to serve many different purposes and it is one of the goals of this book to acquaint the reader with some of these forms. Of course, within the compass of a small work it is possible only to give selected examples of various types of maps, but this selection includes a number of landmark maps in the history of cartography. It was considered better to deal with a limited number of maps in detail than to make an encyclopedic coverage of a larger number without much in-depth discussion. Specific map characteristics will be brought out in reference to particular examples, which are drawn both from historical and contemporary sources. This work is not a treatise on map-making but, rather, one of map appreciation and map intelligence.[2]

This volume may be thought of as a source book of cartographic forms or as an anthology of maps which, like all anthologies, reflects the taste and predilection of the collector. It may also be likened to a book of reproductions of works of art, in the sense that the illustrations, even with the accompanying verbal commentary, cannot really do justice to the originals. In this case, the illustrations are in black and white, many are reduced in scale, and some are merely fragments, reinterpretations, or details. But they will have served their purpose well if people are encouraged by reading this book to look at maps critically, to comprehend their strengths and limitations, to use them more intelligently, and perhaps to collect maps. While no substitute for the map library, it is hoped that this book will lead to the better use of such facilities.[3] Neither

[2] Leading textbooks on modern map-making techniques in English, well-known to American geographers, are: Erwin J. Raisz, *General Cartography,* 2nd ed. (New York: McGraw-Hill, 1948); Arthur H. Robinson and Randall D. Sale, *Elements of Cartography,* 3rd ed. (New York: John Wiley, 1969); and F. J. Monkhouse and H. R. Wilkinson, *Maps and Diagrams,* 2nd ed. (London: Methuen, 1963).

[3] Many countries maintain a map library as part of the national collection. In the United States, the most comprehensive assemblage of maps and atlases, early and modern, foreign and domestic, is contained in the Geography and Map Division of the Library of Congress, while the National Archives is the official depository of U. S. Government maps. The Map Room of the British Museum has exceptionally rich and varied cartographic holdings. Big city libraries often have large map collections as do some government agencies and geographical and other scientific organizations. Larger universities often have important map resources but they are frequently scattered among different departments. However, in some cases, they are centralized as, for example, at the University of California, Los Angeles, where over half a million sheets published since 1900 and aerial photographs are contained in the UCLA Map Library, and maps and atlases published before this date are stored in Special Collections. See Walter W. Ristow, "The Emergence of Maps in Libraries," *Special Libraries,* vol. 58, no. 6 (July–August 1967) 400–419. The Map Section of the Special Libraries Association

is this work a substitute for the rich professional literature of cartography upon which it draws, but it may lead the serious reader to consult these resources.[4]

Cartography cuts across disciplinary lines to a greater extent than most subjects. No one person or area of study is capable of embracing the whole field; and cartographers, like workers in other activities, are becoming more and more specialized with the advantages and disadvantages which this inevitably brings. "Nevertheless," as Hartshorne asserts, "workers in other fields commonly concede without question that the geographer is an expert on maps. . . . This is the one technique on which they most often come to him for assistance. . . ."[5] Accordingly, it is incumbent upon all geographers to understand something about cartography as well as the particular branch of geography in which they specialize. In spite of notable contributions to cartography of a few, many geographers do not possess enough knowledge of maps to serve as advisors to those in other fields who may consult them. This book is intended to help fill this need as well as to promote the use and enjoyment of maps.[6] It is written especially for the non-cartographer who wishes or needs to know something of maps.

---

has done much to promote cartography. An international center with the world's cartographic records on film or tape has been suggested but has not yet been realized.

4 A number of periodicals, especially geography journals, occasionally include articles on map topics. In addition, there are several national and international journals concerned exclusively with cartography. Among the latter group are: *The International Yearbook of Cartography* (Gütersloh: Bertelsmann Verlag), founded by Eduard Imhof in 1961 and published annually since that date, which is the organ of the International Cartographic Association; and *Imago Mundi: A Periodical Review of Early Cartography* (published irregularly in various cities, by various publishers), which was founded in 1935 by Leo Bagrow. See also, Chauncy D. Harris and Jerome D. Fellman, *International List of Geographical Serials* (University of Chicago, Department of Geography Research Paper no. 63, 1960), and Chauncy D. Harris, *Annotated World List of Selected Current Geographical Serials* (University of Chicago, Department of Geography Research Paper no. 96, 1964), for further information on this matter.

5 Richard Hartshorne, *The Nature of Geography,* reprinted with corrections (Lancaster, Pa.: Association of American Geographers, 1967), pp. 247–48.

6 A great many people who have no professional concern with cartography are interested in maps, especially older decorative maps. To help satisfy an increasing demand for these, plates are removed from old atlases and mounted or framed. In addition, a new cartographic genre has arisen—the newly drawn map of antique appearance. A further development, the production of facsimile atlases and maps, is discussed in Walter W. Ristow, "Recent Facsimile Maps and Atlases," *The Quarterly Journal of the Library of Congress* (July 1967), pp. 213–99. Ronald V. Tooley founded The Map Collector's Circle in England in 1963, to cultivate this interest in old maps. See also, Arthur H. Robinson, "The Potential Contribution of Cartography in Liberal Education," *Geography in Undergraduate Liberal Education* (Washington: Association of American Geographers, 1965), pp. 34–47.

One of the main themes of this book is that the modern map can be well-designed, even a thing of beauty and elegance, and that earlier workers had no monopoly on this aspect of cartography. Moreover, the view is here taken that, contrary to the opinion of some, the study of cartography has become increasingly exciting in the last century and a half through the application of modern technology. In recent years this medium of communication has been enriched by new data and is capable of conveying its messages in increasingly precise ways. At the same time, the visual qualities have been vastly improved through the development of new techniques, materials, and processes. However, in the present work some aspects of contemporary cartography are purposely omitted or treated more briefly than their importance suggests that they should be because it is felt that individual books are needed to cover such topics as map transformations, computer mapping, gravity models, etc. The emphasis here will be on landmarks of geo-cartography.

Of course mapping is not confined to the representation of the earth; other phenomena such as the human brain have been mapped. The principles and methods of cartography have a universality that makes them applicable to the mapping of extraterrestrial, as well as terrestrial, bodies. In particular, lunar mapping, which is not a new activity, will receive some attention here. But the main emphasis will be upon what is called geo-cartography. This term will find greater use as we receive and process more detailed information of bodies other than the earth. As space technology develops, it may be desirable to distinguish between extraterrestrial and terrestrial mapping as we now distinguish between astronomy and geography.

A *geographical map* is a representation of all or a part of the earth, drawn to scale, usually on a plane surface. A wide variety of materials have been used in cartography, including stone, wood, metal, parchment, cloth, and paper. The words, map and chart, appear to derive from the materials: the Latin word *charta* denotes paper and *mappa* indicates cloth. In geography today, the term chart is most often applied to maps of the sea, or at least to maps used by sailors and airmen. Map is a broader term in modern usage, and refers more particularly to a representation of land.

To illustrate some of the foregoing ideas, let us examine the cartographic works of primitive societies. That pre-literate peoples, without apparent influence from the outside, engage in mapping attests to the basic importance of cartography to man. Certain Eskimo and American Indian groups, for example, with rudimentary equipment have produced charts which are well suited to their needs, and which compare favorably with those of the same areas made by surveyors of technologically advanced countries. Similarly, nomads in the desert and other peoples whose livelihood, or very existence, depends on their knowledge of particular areas, make sketch maps of these localities with whatever materials are available.[7]

The stick charts of the Marshall Islands illustrate what might be called "native" cartography. Long before Westerners reached the Pacific, stick charts were used by the Marshall Islanders; references to native methods of navigation appear in the European literature of the early nineteenth century. But it was not until the end of the last century that people beyond the Pacific gained a detailed understanding of Marshallese stick charts. The charts of the Marshall Island navigators are made, generally, of narrow strips of the center ribs of palm leaves lashed together with cord made from locally grown fiber plants. The arrangement of the sticks indicates the pattern of swells or wave masses caused by winds, rather than of currents, as was formerly thought to be the case. The positions of islands are marked approximately by shells (often cowries) or coral.[8] The charts vary considerably in size but are usually between 18 and 24 inches square.

The method of using these charts was elicited from the natives with considerable difficulty because their navigational methods were closely guarded secrets. Distances between the various islands of the Marshall group are not great, but because they are low atolls, they can be seen only from a few miles away from an outrigger canoe. To locate an island that he is unable to see, the native navigator observes the relationship between the main waves driven by the trade winds, and the

[7] Among the different media and techniques used for cartographic purposes by earlier and primitive people are: wood, including driftwood, carved to represent relief; wooden boards, bark, skins, leather, and fabric, painted with natural dyes including blood; metal, stone, and clay marked with instruments; clay, sand, and even snow modelled or marked with the hands. Map-making activities of primitive peoples of different geographical milieu have been discussed in literature. A collection of native cartography assembled in Russia in the early years of this century included 55 maps from Asia, 15 from America, 3 from Africa, 40 from Australia and Oceania, and 2 from the East Indies, as indicated in B. F. Adler, *Maps of Primitive Peoples* (St. Petersburg: Karty Piervobytnyh Narodov 1910), and discussed in H. De Hutorowicz, "Maps of Primitive Peoples," *Bulletin of the American Geographical Society,* vol. 43 (1911), 669–79. See also, Clara E. Le Gear, "Map Making by Primitive Peoples," *Special Libraries,* vol. 35, no. 3 (March 1944), 79–83; and Robert J. Flaherty, "The Belcher Islands of Hudson Bay: Their Discovery and Exploration," *Geographical Review,* vol. 5, no. 6 (June 1918), 433–43. Among more civilized peoples, a variety of materials have also been used in cartography. Mosaic maps of tiles, maps woven as carpets or tapestries, or painted mural decorations have been made, as well as globes in the form of goblets and saltcellars.

[8] Marshallese stick charts have been the subject of several articles including, Sir Henry Lyons, "Sailing Charts of the Marshall Islanders," *Geographical Journal,* vol. 72, no. 4 (October 1928), 325–28; and William Davenport, "Marshall Island Navigation Charts," *Imago Mundi,* vol. 15 (1960), 19–26.

secondary waves (reflecting or converging) resulting from the presence of an island. If a certain angle exists between the two sets of waves, a choppy interference pattern is established. When such a zone is reached, the canoe is placed parallel to this feature with the prow in the direction of waves of greater amplitude, which give a landward indication. These often complex wave patterns can be illustrated on the stick charts, which may be carried on the canoe. In addition, the navigators lie down in their craft to feel the effect of the waves.

Three major types of charts are found in the Marshall Islands, namely, *rebbelib, meddo,* and *mattang* (Figs. 1.1, 1.2, and 1.3). The *rebbelib* (Fig. 1.1) is a chart of a large part of the Marshall group, which consists of about thirty atolls and single islands over a distance of approximately 600 sea miles northwest-southeast, and about half that distance northeast-southwest. Although the spatial relationships between the islands are only approximated on the stick charts, these locations can be recog-

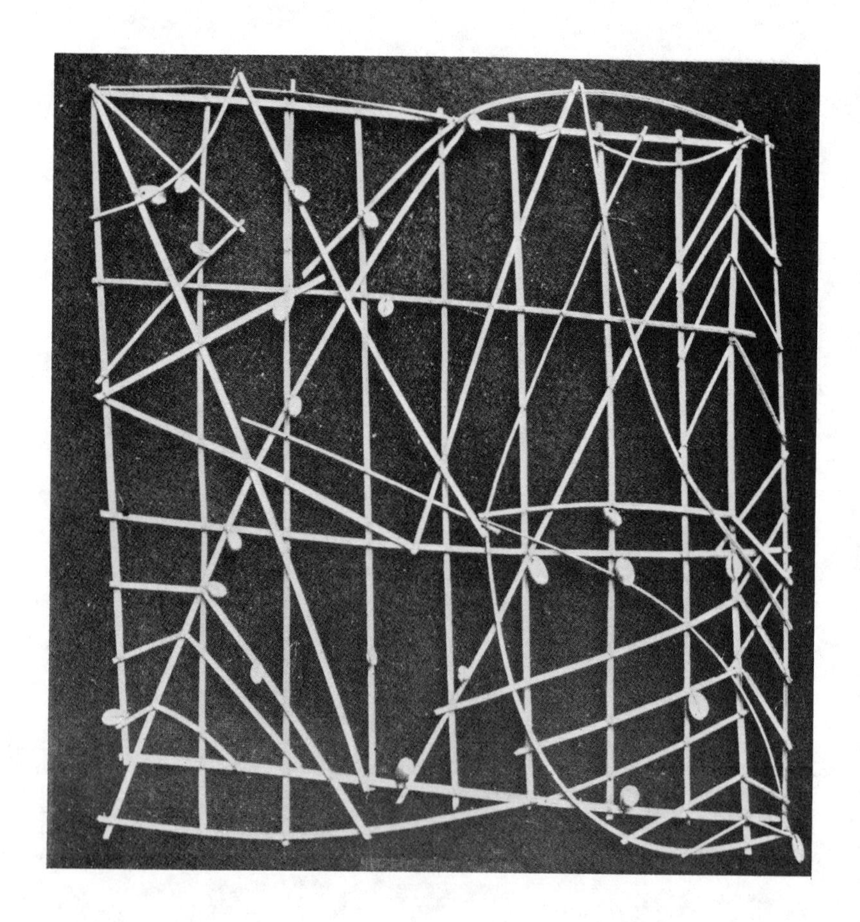

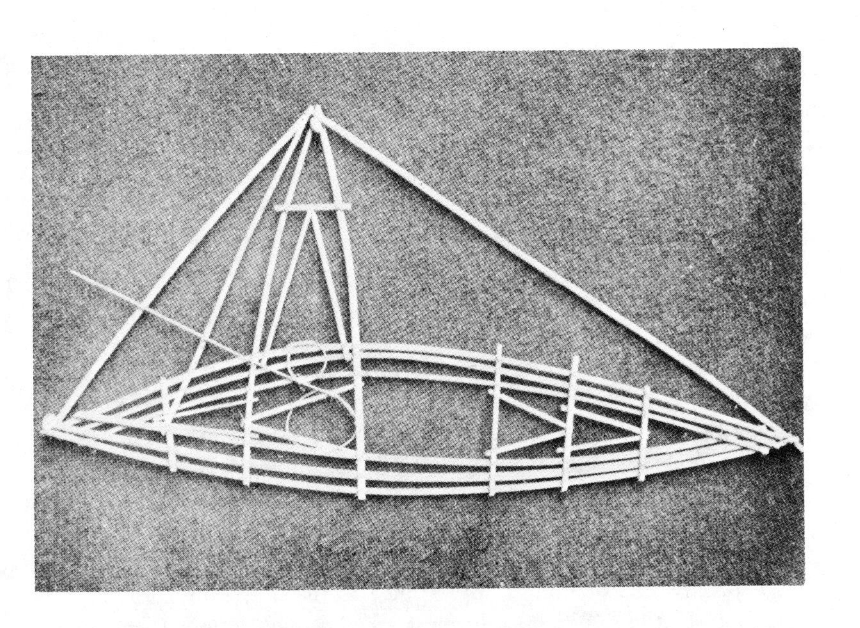

FIGS. 1.1, 1.2, *and* 1.3. *Stick charts from the Marshall Islands.* 1 (opposite) rebbelib *or general chart;* 2 (this page, top) meddo *or sectional chart;* 3 (bottom) mattang *or instructional chart.*

nized by referring to modern navigational charts of the area. The *meddo* (Fig. 1.2) is a sectional chart of part of the island group. It may be one of a series of charts and, because of its scale, allows more detail to be shown than is possible on the *rebbelib*. A third type of Marshallese stick chart, the *mattang* (Fig. 1.3), unlike the others, is not carried on the canoes, but is used for instructional purposes. A *mattang* is a highly conventionalized, often symmetrical chart which does not necessarily show an actual geographical location. It provides a summary of information on wave patterns that might have wide application, although a full understanding of its characteristics may be possessed only by its maker.

In the Marshallese stick charts we can see ingenious, independent, and spontaneous solutions to various cartographic problems. The materials from which they are made (palm, shells, etc.) are available within the limited material resource base of the Islanders, reminding us that not all maps are documents printed on paper. The stick charts illustrate spatial phenomena of infinite importance to the native inter-island navigator, but of little significance to most other people. *Rebbelib* and *meddo* charts indicate the need for maps of different scales; the *rebbelib* shows a broad area and uses a small scale, while the *meddo* illustrates a more restricted locality on a larger scale. In the use of the *mattang,* we recognize the necessity of learning to read charts or maps to understand the relationship between cartographic convention and reality. Furthermore, the desire to conceal a body of geographical information and its cartographic expression, as exemplified by the Marshall Islanders, is a recurring and continuing theme in the history of map-making.

Maps of Antiquity

2

We have noted that primitive peoples of our own time, or close to our time, use widely different means to express themselves cartographically. Similarly, maps by peoples of earlier ages are marked by variety in objective, symbolism, scale, and materials. Understandably, only a small fraction of the maps produced in earlier ages have survived, but in some instances we know of lost works through written records. The loss of many early maps can be attributed to the nature of the materials used for their construction, which often militated against their preservation. Thus, valuable metal was melted down and parchment scraped to be used for some other purpose. Alternatively, less durable materials quickly deteriorated, especially when taken to a different climate, or were destroyed by war, fire, or other means. The destruction of maps is a continuing problem, especially because the information they contain may quickly go out of date so that they are treated as ephemera, or because they have data of a strategic nature which is not to be disseminated.

It is assumed that cartography, like art, pre-dates writing; like pictures, map symbols are apt to be more universally understood than verbal or written ones. Maps produced by contemporary primitive peoples have been likened to so-called prehistoric maps. Certain carvings on bone and petroglyphs have been identified as prehistoric route maps, although according to a strict definition, they might not qualify as maps. Some of these carvings were probably made as a pastime but others may have been used to indicate hunting and gathering sites. As Carl Sauer has suggested, such maps might show "the route to the fat oysters" and, in turn, a place may have received its name from the association.[1]

To illustrate the cartography of pre-literate man, we chose, from a number of possibilities, the navigation charts of the Marshall Islanders. Likewise, to exemplify the cartography of literate peoples in antiquity, we will emphasize principally the maps of one area, Mesopotamia. However, a map from ancient Egypt is included because it represents a type of which a number of examples have survived.[2] This particular map (Fig. 2.1), which portrays the route to Paradise, takes its inspiration from the valley of the Nile. The original is painted, in black, red, and white with an ochre background, on the bottom of a wooden sarcophagus from the Middle Empire. This highly stylized rendering may be considered as a very early example of a *theoretical map* or model, a persistent form in cartography in spite of the admonitions of some that geographers should address themselves only to the real world. Other early maps from Egypt, and also from Mexico, use realistic pictorial devices rather than diagrammatic symbols as shown in Figure 2.1. It should be mentioned that the ancient Egyptians are credited with the invention of geometry, in response to the need for property surveys and re-surveys as the flooding of the Nile erased boundary markers, and it is believed that cadastral or property maps were made by these people. However, the only known survivors of this type of map are plans of various buildings and a map of the Nubian gold mines. The latter was drawn on a roll of papyrus and shows, in addition to the mining site, a road, a temple and a few other buildings.

The earliest map by civilized man that we know about comes from Mesopotamia. To illustrate the diversity of scale and purpose of the cartography of this area, we reproduce three examples (Figs. 2.2, 2.3, and 2.4). Although these maps are different in several particulars, they are alike in that all of them were drawn with cuneiform characters and stylized symbols impressed or scratched on clay tablets. (This method placed great restrictions on the cartographer, for a series of straight lines are frequently employed to approximate curves. This device makes some of these early maps strangely reminiscent of some modern computer

[1] Carl O. Sauer in the banquet address of the Spring Meeting of the California Council for Geographic Education held at California State College at Hayward, California on May 4th, 1968. Professor Sauer devoted much of his address to the importance of the map in geographical studies.

[2] The map is from Prince Youssouf Kamal, *Monumenta Cartographica Africae et Aegypti,* 5 vols. in 15 fascicules (Cairo, 1926–51). This greatest of facsimile atlases, like a number of other such works, was distributed in a very limited number of sets. It is important not only for the study of Africa, but also to illustrate changing cartographic forms from ancient Egypt to the period of modern exploration. See Norman J. W. Thrower, "Monumenta Cartographica Africæ et Aegypti," *UCLA Librarian,* suppl. to vol. 16, no. 15 (1963), 121–26.

FIG. 2.1. *Early Egyptian map from a wooden sarcophagus.*

cartography in which the constraints of the machine often force us to approximate curving lines with straight line segments.) We will consider these three Mesopotamian maps in terms of scale, dealing first with that of largest scale which indicates a small area in the Nippur district.[3] This

[3] Ekhard Ungar, "Ancient Babylonian Maps and Plans," *Antiquity,* vol. 9 (1935), 311–22; and "From Cosmos Picture to World Maps," *Imago Mundi,* vol. 2 (1937), 1–7. See also, Theophile J. Meek, "The Orientation of Babylonian Maps," *Antiquity,* vol. 10 (1936), 223–26.

FIG. 2.2. *Early Mesopotamian city plan on a clay tablet.*

fragment (Fig. 2.2) shows and identifies large and smaller canals, a city wall with gates and a moat, houses and their openings, a terrace, etc.

A Mesopotamian map of intermediate scale, though of such small dimensions that it can be held in the palm of the hand, is the well-known Akkadian map found at Nuzi and dated c. 2500 B.C. (Fig. 2.3), sometimes described as the oldest map in the world. This map is oriented with east at the top and certain features can be clearly identified. These include water bodies, settlements, and mountains. The latter are shown by fish scale-like symbols at the top and bottom of the map, an atypical form of representation, for the plan view is used for other features and characterizes Mesopotamian cartography generally.

The third in this series (Fig. 2.4) is a world map with an Assyrocentric view and, naturally, is of smaller scale than the other examples. It shows a round, though presumably flat earth with Babylon in the center. The Euphrates flows from its source in the Armenian mountains

FIG. 2.3. *Early map of Mesopotamia on a clay tablet.*

FIG. 2.4. *Early Mesopotamian world map on a clay tablet.*

in the north to the Persian Gulf where it joins an encircling sea. Indeed, the purpose of the map seems to be to show the relationship between the "Earthly Ocean," represented by a circle, and the "Seven Islands" illustrated by triangles (only one of which is entirely intact on the tablet). We need not explain in detail the astrological and religious significance of this map, which is indicated in the text on the tablet, but we can make some general statements about it. Understandably other civilizations, including the Chinese, have taken an egocentric view of the world; the long-held geocentric theory may be the ultimate expression of such an idea applied to the universe. Looking forward, we may note here that the separation between the terrestrial and the celestial spheres was an important theme in the Middle Ages in Europe, as was the concept of a circumfluent ocean. In considering the circle in cartography, it should be mentioned that the sexagesimal system of dividing this figure, which is the usual method employed in mapping to this day, came to us from Babylon by way of Greece.

Maps that are known only through descriptions or references in the literature (having either perished or disappeared) are a problem to historians of cartography. In some instances, later reconstructions are included in the correct chronological order of the original maps and this can be amply justified when a particular area or civilization forms the main basis of discussion.[4] In the present work, however, the principal emphasis is upon map representation, so that it is difficult to take this approach. All reconstructions are, to a greater or lesser degree, the product of the compiler and the technology of his times. Therefore, reconstructions will be used here only to illustrate the cartography of the period in which the particular map was made. Nevertheless, reconstructions of maps which are known to have existed, and which have been made a long time after the missing originals, can be of great interest and utility to scholars. The possibilities include those for which specific information is available to the compiler and those which are described or merely referred to in the literature. Of a different order, but also of interest, are those maps made in comparatively recent times that are

[4] Maps of many periods and schools, focusing on a particular area, are contained in *Monumenta Cartographica Africæ et Aegypti;* in this work, reconstructions of maps no longer extant are used in place of originals or assumed originals. The reconstructions of such maps appear in the correct chronology of the originals irrespective of the date of the reconstruction. Examples of reconstructed maps focusing on Greco-Roman civilization are found in J. O. Thomson, *Everyman's Classical Atlas* (London: J. M. Dent, 1961). This readily available work contains maps of the world according to Hecatæus, Herodotus, Eratosthenes, Crates, and Ptolemy, some of which are derived from Sir Edward Bunbury, *A History of Ancient Geography,* 2 vols. (London: John Murray, 1883), a work of fundamental importance on the geographical and cartographical knowledge of the period.

designed to illustrate the geographical ideas of a particular person or group in the past but are suggested by no known maps.[5]

The lack of direct cartographic evidence has made consideration of early Greek mapping somewhat speculative. Recently a series of Greek coins was found, dating from the fourth century B.C. with images, one of which is claimed to be "the earliest Greek map to come down to us in any form and the first physical relief map known."[6] However, scholars have often used reconstructions of maps to illustrate the cartography of antiquity. Although not used as illustrations in this section of the work, reconstructions of, or at least maps which owe their inspiration to those of the Greeks and Romans will be used in Chapters 4 and 5; here we will deal briefly with some contributions of these people to cartography and geography. There was rarely a clear distinction between these two disciplines in antiquity, and a number of workers who were designated geographers were, in fact, cartographers. However, there was a sharp dichotomy in ancient Greece between land measurers (geometers), who were employed to delineate small areas, and philosophers, who speculated on the nature and form of the entire earth. The idea of the earth as a slab eventually gave way to a drum or pillar-shaped world favored by the Ionian, Anaximander of Miletus who, as early as c. 550 B.C., drew a world map. This was improved upon fifty years later by Hecatæus, also from Miletus, who believed that the upper surface of the earth pillar was a curving disc. The Mediterranean was in the center of a world island composed of the lands bordering this sea, the whole being surrounded by a circumfluent ocean (ocean-stream). It is assumed that the Gnomonic projection (see Appendix A) was developed by the early Ionian philosophers, but that it was used only for astronomical purposes at this time. There was much discussion among the Ancients concerning the major divisions of the land area from which arose the concept of the three continents of the Old World. Through his writings of travels, Herodotus (active 440–425 B.C.) did much to enlarge contemporary knowledge of Asia. The travels in the east of Alexander the Great (died 323 B.C.), and those of his contemporary Pytheas in the north, further enlarged the world known to the Greeks. Meanwhile an idea fundamental to later progress in cartography, the spherical form of the earth (which probably had its beginnings among the Pythagoreans), was gaining currency with the Platonic philosophers.

[5] In addition, purely imaginary maps have been drawn to illustrate novels and other literary works as well as those which are suggested by known landscapes; an example of the first kind would be the Hobbit maps of J. R. R. Tolkien and, of the second, Thomas Hardy's Wessex maps.

[6] A. E. M. Johnston, "The Earliest Preserved Greek Maps: A New Ionian Coin Type," *Journal of Hellenic Studies,* vol. 87 (1967), 86–94.

A scholar with vision large enough to put this information into a logical framework was needed, and such appeared in the person of Eratosthenes, (276–196 B.C.). Eratosthenes, head of the Library at Alexandria from 240 B.C. until his death, was known as *beta* to his contemporaries because they considered him second in all his varied academic pursuits. More critical of these accomplishments was Strabo (63 B.C.?–A.D. 24) to whom we are indebted for much of our knowledge of geography in antiquity, including the work of Eratosthenes. Later workers have a higher opinion of Eratosthenes, regarding him, "as the parent of scientific geography"[7] and at least "worthy of *alpha*"[8] in that subject, particularly for his remarkable measurement of the circumference of the earth. Once the idea of a spherical earth was accepted, the measurement of this body was a logical step—even to Greek scholars who were more given to philosophical speculation than to quantification and experimentation. Eratosthenes was not the first to compute a figure for the circumference of the earth; this distinction may belong to Eudoxus of Cnidus (c. 370 B.C.) who estimated its measurement at 400,000 stades. A figure of 300,000 stades is credited to Dicæarchus (died 296 B.C.), a student of Aristotle. A similar figure was proposed by Aristarchus of Samos (died 230 B.C.), who has been called the Copernicus of Antiquity because of his early espousal of a heliocentric rather than geocentric view of the universe. Perhaps, more properly, Copernicus should be called the Renaissance Aristarchus.

Both the method and the accuracy of Eratosthenes' well-known measurement of the earth have evoked the admiration of later workers, and this calculation is regarded as one of the great achievements of Greek science. Eratosthenes observed that the rays of the sun, at midday, at the time of the summer solstice, fell directly over Syene (Aswan) and that the vertical rod of the sun dial (gnomon or style) would not cast a shadow. At the same time of day and year, the shadow cast by a gnomon at Alexandria, to the north of Syene, was measured by Eratosthenes as 1/50 of a proper circle (Fig. 2.5). He assumed that: Syene (S) and Alexandria (A) lie under the same meridian circle; that rays (R^1 and R^2) sent down from the sun are parallel; that straight lines falling on parallel lines make alternate angles equal; and that arcs subtended by equal angles are similar (angle ACB is equal to angle SZA). He accepted a figure of 5,000 stades for the distance from Syene to Alexandria, which, according to his previous reasoning, was 1/50 of the circumference of the earth. Thus 5,000 stades × 50 equals 250,000 stades

[7] Bunbury, *A History*, vol. 1, p. 615.
[8] Walter W. Hyde, *Ancient Greek Mariners* (New York: Oxford University Press, 1947), p. 14*n*.

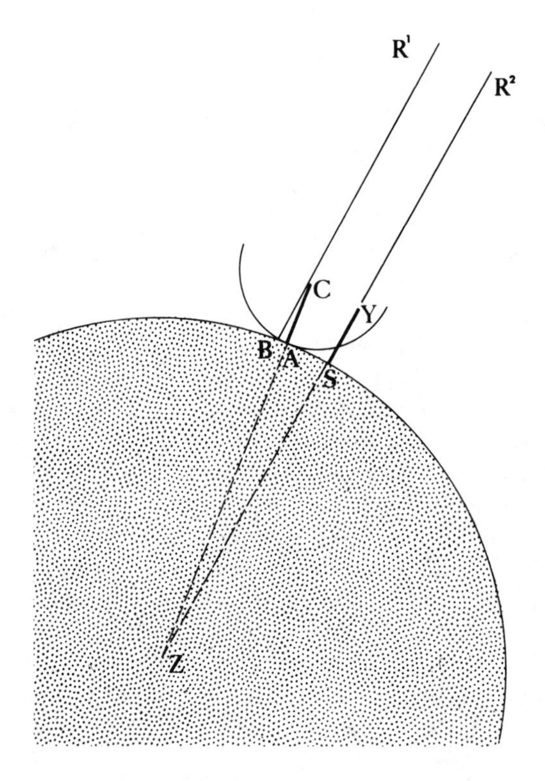

FIG. 2.5. *Diagram of Eratosthenes' measurement of the circumference of the earth.*

for the whole circumference of an assumed perfectly spherical earth. There has been much debate about the length of the stade[9] and we know that Syene is approximately 3 degrees east of Alexandria and some 37 miles north of the tropic. Nevertheless, it is now thought that Eratosthenes' measurement may have been within 200 miles of the correct figure of the circumference of the earth.

Not content with this achievement, Eratosthenes also attempted to divide the earth in a meaningful way. In this, too, he followed Dicæarchus who had separated the known inhabited world into northern and southern parts with a line passing through the Pillars of Hercules (Strait of Gibraltar), eastward to what is now Iran. Eratosthenes accepted this line and drew another at right angles to it passing through Rhodes, where the two lines intersected. In these developments, and in the addi-

9 See Jacob Skop, "The Stade of the Ancient Greeks," *Surveying and Mapping,* vol. 10, no. 1 (1950), 50–55.

tion of unequal zones parallel to the first line of latitude, we can see the beginnings of an earth grid.

Other accomplishments of Eratosthenes affecting cartography were his measurement of various distances on the earth's surface, including the length of the Mediterranean (the best for the following 1300 years), and his considerable additions to the world map, especially in southern Asia and northern Europe, made possible through information provided by travelers. Eratosthenes also argued for a predominantly water-covered earth in contrast to others, such as Crates, who proposed a largely terrestrial sphere. Crates of Mellos (active second century B.C.) conceived and may have constructed, a globe delineating four approximately symmetrical continents—two in the northern hemisphere and two in the southern, separated from each other by relatively narrow (circumfluent) bodies of water. This teleological idea was in opposition to the scientific view of Eratosthenes, but Crates' concepts and similar ideas of others persisted for centuries. A contemporary of Crates, the famous astronomer Hipparchus of Nicæa, accepted Eratosthenes' measurement of the globe which he divided by a systematic, imaginary grid of equally spaced parallels (*klimata*) and meridians crossing each other at right angles. Although a few points were astronomically fixed, positions of places on the earth were largely estimated because of the difficulty of measuring longitude, especially. Hipparchus, who used 360 degrees of 700 stades each for the earth's circumference, also insisted on the accurate location of places according to latitude and longitude, as determined by astronomical observation. Once a systematic earth grid was adopted, the serious study of map projections was possible—in fact, the Stereographic and Orthographic projections, which have become popular in recent years and which will be referred to later, are attributed to Hipparchus. However, like the Gnomonic, they were probably used only for astronomical purposes at this time.

In the first century A.D., Marinos of Tyre (Marinus), like Eratosthenes before him, attempted to enrich the world map by incorporating information from new sources. Marinos also devised a simple rectangular (plane) chart based on the latitude of Rhodes. These contributions influenced Ptolemy, who was a later critic of Marinos, to greater accomplishments in cartography. Ptolemy (Klaudios Ptolemaios), who flourished in the second century A.D., was Librarian of Alexandria, the position held some four centuries before by Eratosthenes. Like his predecessor, Ptolemy took giant strides in various phases of cartography which were not materially improved upon for many centuries after his death.[10] Ptolemy

10 See Leo Bagrow, "The Origin of Ptolemy's 'Geographia,'" *Geografiska Annaler,* vol. 27, no. 3–4 (1945), 318–87.

did not use Eratosthenes' measurement of the earth, but rather employed the smaller measure of the Greek astronomer Poseidonius (186–135 B.C.). This "corrected" figure of the circumference of the earth, some 180,000 stades or roughly three-quarters of the actual distance, had also been used by Marinos and others. Because of its adoption by Ptolemy, whose authority as both astronomer and cartographer was not seriously challenged for fourteen centuries following, the error was perpetuated. Nevertheless, Ptolemy's specific contributions to cartography were of the greatest importance. They are contained, along with material from other scholars, in his guide to making maps, a work now known as the *Geographia* or simply the *Geography*. Ptolemy's *Geographia* included: (*1*) instructions for making map projections of the world (simple conic and spherical, with concentric arcs); (*2*) suggestions for breaking down the world map into larger scale sectional maps (in some editions there are twelve of these for Asia, ten for Europe, and four for Africa, while others propose a much larger number of regional maps); (*3*) a list of coordinates of some 8,000 places. For the latter, two different systems were employed: i.e., latitude and longitude in degrees; and latitude according to the length of the longest day and longitude in time (1 hour = 15 degrees) from a prime meridian. Ptolemy's prime meridian (0 degrees longitude) passed through the Fortunate (Canary) Islands and his map extended 180 degrees eastward to China (see Figs. 5.1 and 5.2). Although, like Hipparchus, Ptolemy argued for the astronomical determination of earth locations, in fact most of those contained in the *Geographia* were supplied by travelers and were based on dead reckoning (position from courses sailed and distances made on each course).

Ptolemy's work has come down to us through later copies preserved in the Byzantine Empire. From these manuscripts, some of which contain maps, it has been possible for scholars to reconstruct knowledge of the world in the centuries immediately following the birth of Christ. As we shall see later, Ptolemy was the ultimate authority on cartography at the beginning of the great European overseas geographical discoveries of the Renaissance. Perhaps it is incorrect to attribute all these developments specifically to Ptolemy and better to think of a Ptolemaic corpus—similar to the Hippocratic tradition in medicine—to which a number of workers contributed.

Cæsar controlled the Alexandria of Ptolemy, who, like Strabo before him and other later Greek scholars, labored for Roman masters. Through such means, the Romans became the heirs of the geographical knowledge of the Greeks which, as we have seen, included: the idea of a spherical earth; measurements of the circumference of the earth; irregular and regular divisions of the sphere (coordinate systems); map projections; maps of different scales; a world map which embraced large parts of

Europe, Africa, and Asia and which, understandably, was progressively less accurate as distance from the Mediterranean increased. From the available evidence, the Romans appear to have been eminently practical in their own cartographic work, being concerned with maps to assist in the military, administrative, and other concerns of the Empire. We know of such Roman maps through references in literature which indicate that, in addition to archival copies on linen, stone or metal duplicates were made for public display. We also know of their existence through later copies, especially the well-known *Tabula Peutingeriana* (Peutinger Table), which will be discussed in connection with the cartography of the Middle Ages. It is sufficient to say that the Roman interest in the orderly layout of lands for settlement and building of roads suggests that cartography was important to them.[11] Except for a few sketch maps, cadastral plans (some on stone), the representation of the Black Sea on the shield of a Roman soldier (probably Greek work), and copies as indicated above, however, we have little direct knowledge of Roman mapping. We can now turn to the Orient where important developments affecting cartography were taking place.

11 Roman Centuriation or rectangular division of land and its enduring effects are discussed in John Bradford, *Ancient Landscapes* (London: G. Bell, 1957), pp. 145–216, and in George Kish, "Centuriatio: The Roman Rectangular Land Survey," *Surveying and Mapping*, vol. 22, no. 2 (1962), 233–44. A Roman surveying instrument found at Pompeii is discussed in Don Gelasio Caetini, "The 'Groma' or Cross Bar of the Roman Surveyor," *Engineering and Mining Journal-Press* (November 29, 1924), p. 855; see also, D. A. W. Dilke, "Illustrations from Roman Surveyors' Manuals," *Imago Mundi*, vol. 21 (1967), 9–29.

The record of early cartography in the East is as fragmentary as it is in the West. Maps drawn in ancient India on palm leaves or on material made of palm fiber, for example, are not extant although it has been postulated that such were probably made.[1] There appears to have been contact between India and Mesopotamia at an early date, and influences, including mathematical and cosmological ideas, traveled in both directions. Similarly, at a somewhat later period, there were religious, scientific, and other cultural exchanges between India and China. It is appropriate to review briefly the early geographical and cartographical contributions of the Chinese both because they are important in their own right and because they exerted an influence on other parts of the Orient. We will use examples from China, particularly, to illustrate the cartography of eastern Asia, as we used examples from Mesopotamia to exemplify early mapping of the Middle East.[2]

In the literature of China, we have evidence of geographic and cartographic activity of a much earlier

[1] Leo Bagrow, *History of Cartography,* revised and enlarged by R. A. Skelton (Cambridge, Mass.: Harvard University Press, 1964), p. 207. This also appeared in a German edition under the title *Meister der Kartographie* (Berlin: Safari-Verlag, 1963). These works are based on Leo Bagrow, *Geschichte der Kartographie* (Berlin: Safari-Verlag, 1943). The Bagrow-Skelton book deals with the history of cartography to the 18th century, and is clearly the best one volume general work on the subject up to that date.

[2] The most readily available English account of the cartography of China is in Joseph Needham and Wang Ling, "Mathematics and the Sciences of the Heavens and the Earth," *Science and Civilization in China,* vol. 3 (Cambridge: Cambridge University Press, 1959), pp. 497–590. See also, E. Chavannes, "Les deux plus ancient specimens de la cartographie chinoise," *Bulletin de l'Ecole Françoise d'Extreme Orient,* 3 (1903).

Early Oriental Cartography

3

date than that of the oldest surviving maps of this civilization. The earliest Survey of China (Yü Kung) is approximately contemporaneous with the earliest reported map-making activity of the Greeks—that of Anaximander (sixth century B.C.). In the centuries following, there are remarkable parallels between the geographical literature of China and that of Greece and the Latin West—suggesting, in some instances, more than casual contacts between these two cultures. As in the West, so in China, we can identify: anthropogeography; descriptions of the home area and of foreign countries; coastal and hydrographic books; urban as well as local topographic studies; and geographical encyclopedias.

It is assumed that maps, charts, and plans accompanied even very early examples of these geographical works. A specific reference in Chinese literature alludes to a map painted on silk in the third century B.C.—the weft and woof of the material may, in fact, have suggested a map grid. We also learn of various rulers, generals, and scholars during the Han Dynasty (207 B.C.–A.D. 220) having a high regard for maps and using them for military and administrative purposes. Apparently the rectangular grid (a coordinate system of equal squares), which is basic to much scientific cartography in China, was formally introduced by the astronomer Chang Heng (a contemporary of Ptolemy). The grid subdivides a plane or flat surface; this figure was assumed for purposes of map-making but it must not be supposed that all scholars in China believed that this was the shape of the earth. Indeed, we know that the Chinese used the gnomon and were aware of the continual variation in the length of its shadow in the long north-south extent of their own country, knowledge that presumably suggested to them a curving surface, if not a globe.

In the third century A.D., we learn of a Prime Minister's younger sister embroidering a map to make the record more permanent. In the same century, Phei Hsiu, a Minister of Works during the Chin Dynasty, outlined the principles of official map-making, which included: the rectangular grid for scale and locational reference; orientation; triangulation; and altitude measurement. None of the maps of Phei Hsiu has survived, but modern scholars have attempted reconstructions of these cartographic works from written descriptions.

As the Imperial territories of China increased through the centuries, maps of various scales were made of the enlarged realm. These works set the stage for even more important cartographic accomplishments which may be exemplified by a map, among the oldest surviving in China, dated 1137 A.D. or even earlier. The map in question (Fig. 3.1) has a regular, rectangular grid with the scale of 100 li (about 36 miles) to each square. It delineates the coastline and major rivers of China in

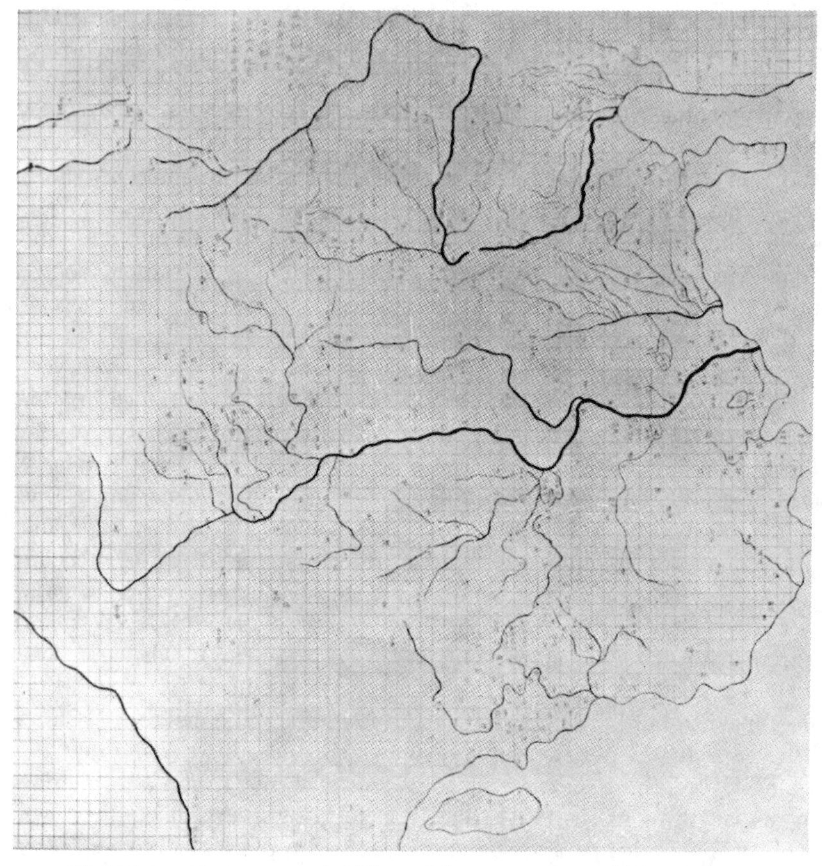

FIG. 3.1. *Early map of China with rectangular grid, carved on stone.*

clearly recognizable form. (An outline map of this large area, with details taken from a modern chart, is provided for comparison in Fig. 3.2). The map in Fig. 3.1, which is about 3 feet square, was carved in stone by an unknown cartographer of the Sung Dynasty; its purpose was to illustrate a much earlier geography, the before mentioned survey, Yü Kung. Needham and Ling assert, with justice, that this map is "the most remarkable cartographic work of its age in any culture."[3] Its portrayal of the coastline and drainage of China is better than that on any map, European or Oriental, until the period of modern systematic surveys.

Another Chinese cartographic milestone of about the same period is the earliest known printed map. It is assumed to have been made

[3] Needham and Ling, *Science and Civilization,* Plate LXXXI, facing p. 548.

FIG. 3.2. *Outline map of China from a modern chart, for comparison with Fig. 3.1.*

around 1155 A.D., and predates the first printed European map by over three centuries. This map (Fig. 3.3), which served as an illustration in an encyclopedia, is printed in black ink on paper (which had been invented in China about a millenium earlier), and shows part of western China. In addition to settlements and rivers, a portion of the Great Wall is indicated at the north. Both this map and the one illustrated earlier (Fig. 3.1) have north orientation, i.e., north is at the top of the map which, of course, is now conventional in the West. It is assumed that the Chinese learned of orientations other than this from different peoples with whom they had contact, such as the Arabs, who settled on the China coast before 750 A.D. and who used south-oriented maps.

The culmination of indigenous Chinese cartography is found in the contributions of Chu Ssu-Pen (1273–1337 A.D.) and his successors, who

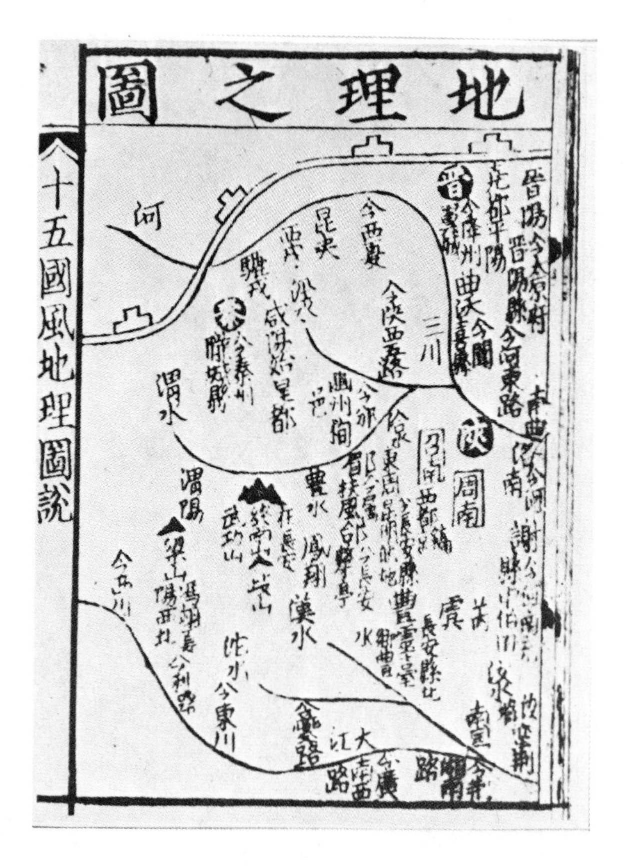

FIG. 3.3. *The earliest printed map, depicting a portion of western China, and showing part of the Great Wall, rivers, mountains, and settlements.*

established a mapping tradition that lasted until the nineteenth century. Chu, who built upon a scientific cartographic heritage extending back to Phei Hsiu and Chang Heng, made a manuscript map of China with a rectangular grid. The reliability of the information on which his map was based was of the greatest concern to Chu, whose attitude is quite modern in this respect. The map was constantly revised, and eventually was enlarged, dissected, and printed in atlas form some two centuries after Chu's death (Fig. 3.4). The treatment of the ocean with angry lines, common in Oriental cartography, perhaps suggests perception of the seas as a hostile environment.

Although this book is more concerned with maps than the methods used to produce them, it should be mentioned that, at least by the time of Chu Ssu-Pen, the Chinese cartographers knew principles of geometry

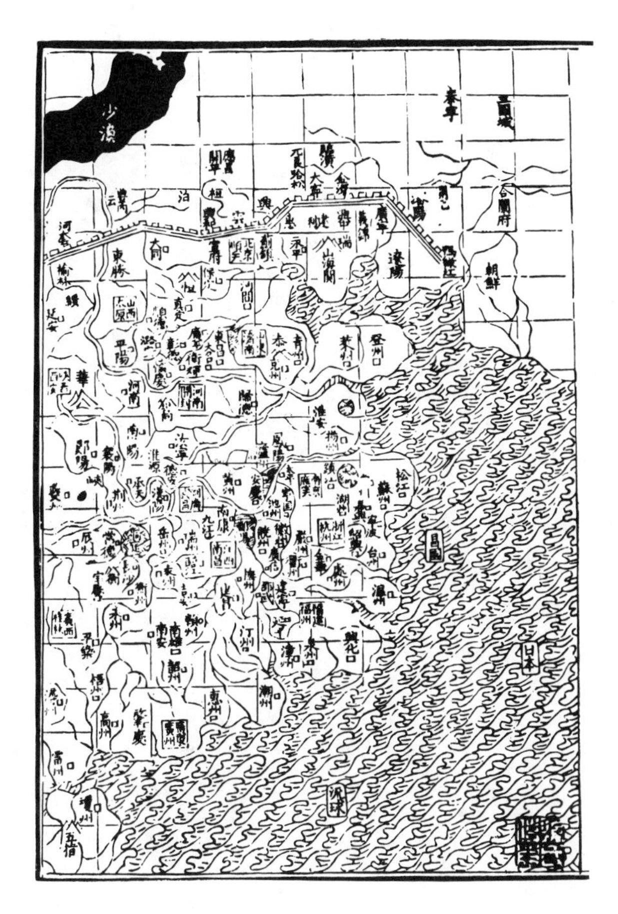

FIG. 3.4. *Atlas map of eastern China, showing parts of the Gobi (solid black area), the Great Wall, and the China Sea.*

and possessed instruments that would greatly facilitate their mapping activities. The instruments included the gnomon, as mentioned previously, and a device similar to the *groma* of the Romans, with plumb lines attached. The Chinese also used sighting tubes and something akin to the European cross-staff for estimating height, as well as poles for leveling and chains and ropes for ground measurement. The odometer or carriage measuring instrument, by which distance is ascertained by the revolutions of the wheels, is referred to in China at least as early as in Europe. Compass bearings, implying the use of the magnetic needle, seem to have been made by the eleventh century A.D.; it is assumed that the magnetic needle was transmitted westward to Europe shortly after this period.[4] Actually, a reference to a south-pointing chariot, which alludes to a mounted magnetized needle, goes back to the Wei Dynasty (third century A.D.), but we are not sure when this instrument was first used for map-making.

The Chinese seem to have made maps of large areas beyond their own borders but, understandably, the quality of the representations decreased with the distance from the culture hearth of the country.[5] We have mentioned that Chinese cartography influenced that of other areas in eastern Asia. This was especially true for Manchuria and Korea, although the cartography of these areas is not without innovation.[6] Japanese cartography seems to have had a more independent development but, again, few really early maps from this area have survived.[7] The early cartography of southeast, southern Asia, and Persia is known to us through references in the literature of these areas. We shall now return to the West, but not before mentioning that when Jesuit fathers in the sixteenth century established residence in China, the cartographic record of this area was made available to the Europeans and incorporated in their world maps. Although after this time Chinese cartography was influenced by European techniques, we do well to recall the extraordinary cartographic accomplishments in the Orient, particularly as exemplified by the works of early Chinese map-makers.

[4] Lynn T. White, *Medieval Technology and Social Change* (Oxford: Oxford University Press, 1962), p. 132.

[5] H. B. Hulbert, "An Ancient Map of the World," *Bulletin of the American Geographical Society,* vol. 36, no. 9 (1904), 600–605.

[6] Norman J. W. Thrower and Young Il Kim, "Dong-Kook-Yu-Ji-Do: A Recently Discovered Manuscript of a Map of Korea," *Imago Mundi,* vol. 21 (1967), 10–20.

[7] The earliest known indigenous Japanese maps are discussed and illustrated in Ryuziro Isida, *Geography of Japan* (Tokyo: Society for International Cultural Relations, 1961), pp. 5–7. Occidental contributions to the mapping of the Orient before the period of modern surveys are considered in Hiroshi Nakamura, *East Asia in Old Maps* (Tokyo: Kasai, 1964). See also George H. Bean, *A List of Japanese Maps of the Tokugawa Era* (Jenkintown, Pa.: Tall Tree Library, 1951, and supplements in 1955, 1958, and 1963).

Links between antiquity and the Middle Ages are provided by several maps made in the latter period, which express certain classical as well as medieval ideas. One such map is the well-known *Tabula Peutingeriana* (Peutinger Table), alluded to in respect to Roman cartography and which takes its name from a sixteenth century owner of the map, the Humanist, Konrad Peutinger of Augsburg.[1] The *Tabula* is a large manuscript map copied on parchment, roughly 1 foot high and with an overall length of more than 20 feet. It originally consisted of twelve sections, the first of which, now lost, is thought to have shown most of Britain and the Iberian peninsula, and adjacent parts of North Africa. It might also have contained the name of the cartographer, who is believed to be Castorius. The surviving eleven sections illustrate, in a highly diagrammatic fashion, the area from southern England through the Mediterranean to India. It is presumed that the information is from Roman itinerary maps of the first century A.D., though some scholars have identified it with the map of Marcus Vipsanius Agrippa (63–12 B.C.). As it has come down to us, the *Tabula* appears to be mainly fifth century work with some additions made as late as the ninth century.

Figure 4.1 shows a small portion of the Peutinger Table, where sections VI and VII adjoin, which focuses on Sicily and the foot of Italy. As can be seen from this sample, areas on the map are very elongated in the east-west direction. The map is not developed on a systematic projection but approximate distances between settlements are written on the map. The *Tabula* is

[1] Konrad Miller, *Die Peutingerische Tafel* (Stuttgart: *Brockhaus,* 1962), pp. 1–16, I–XII and half-scale color reproduction of the map.

4 *Cartography in Europe and Islam in the Middle Ages*

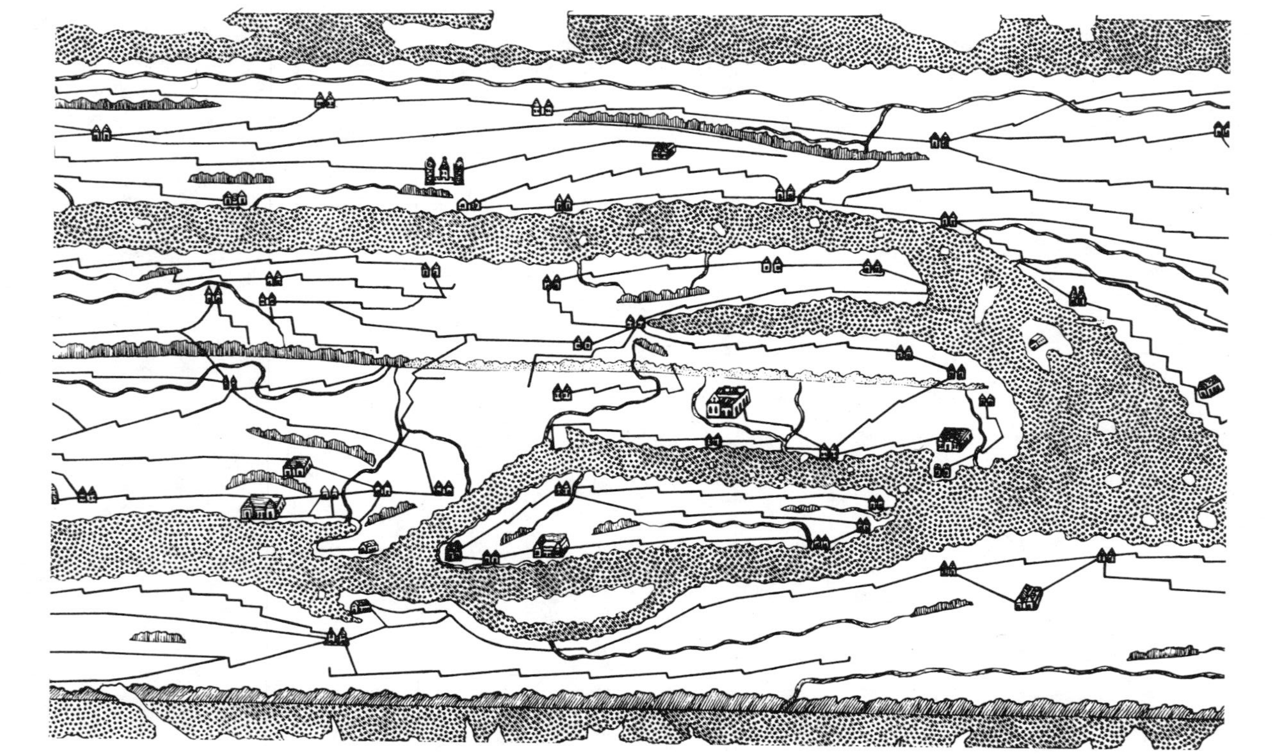

FIG. 4.1. *Simplified rendering of a small section of the Peutinger Table, focusing on the boot of Italy and Sicily. Redrawn by Noel Diaz. Because of the greatly reduced scale of the map, lettering (which forms an important feature of the original) is omitted.*

basically an itinerary or route map with the roads delineated predominantly by straight lines, often with curious jogs; routes are in red, while the sea is indicated in greenish-blue. The map is enriched by the rendering of mountain chains (in profile), buildings and, in some cases, figures at important centers.

For a compilation such as the Peutinger Table, as with early Chinese maps, a spherical earth need not be assumed. This is especially true of maps of small areas such as the city plans of Rome and Constantinople which were part of the library of Charlemagne (742–814). These maps, and also a world map belonging to Charlemagne, were engraved on silver tablets, but we know of them only through references in literature. There is no proof that, in general, medieval man believed in a flat earth. In fact, we know specifically that a number of influential savants of this period accepted a globular world. Nevertheless, there were those who argued against this concept, as well as the related idea of inhabitants in the Antipodes.[2] In the Middle Ages, the world was represented on maps by various shapes: irregular, ovoid, rectangular, cloak-shaped, and circular. The most common shape was the circular, disc, or wheel form, a descendant of Roman cartography, of which two distinct types can be recognized—the T-O, and the climatic zonal forms (Figs. 4.2 and 4.3).

The T-O (*orbis terrarum*) type usually has east at the top (hence the term orientation), with Asia occupying the upper half. Asia is separated from Africa by the Nile and from Europe by the Don (Tanais). Together, these two rivers form the top of the T, while the Mediterranean, which separates Europe and Africa, forms the upright segment of the T. The whole is surrounded by a circumfluent ocean—the O. This cartographic form embodies geographical concepts from antiquity that we have encountered earlier. These include the notions of three continents—a widely accepted idea—and of land covering most of the earth with the separating seas and circumfluent ocean, exemplified by the ideas of Crates. T-O maps also satisfy Christian theology, by giving Jerusalem a central position on the earth.[3]

[2] Among prominent churchmen who apparently accepted the idea of a spherical earth were Adam of Bremen, Albertus Magnus, and Roger Bacon; those who opposed an inhabited Antipodes, if not a globular world, included Lactantius, St. Augustine of Hippo, and the oft-quoted Constantine of Antioch (Cosmos Indocopleustes), author of the *Christian Topography*. General works which discuss medieval cartography in the larger setting of the geography of the times include, John K. Wright, *Geographical Lore at the Time of the Crusades* (New York: Dover, 1965), reprinted; and George H. T. Kimble, *Geography in the Middle Ages* (London: Methuen, 1938).

[3] "This is Jerusalem; I have set her in the center of nations with countries round about her." *Ezekiel* 5:5, Holy Bible Revised Standard Version, 1952.

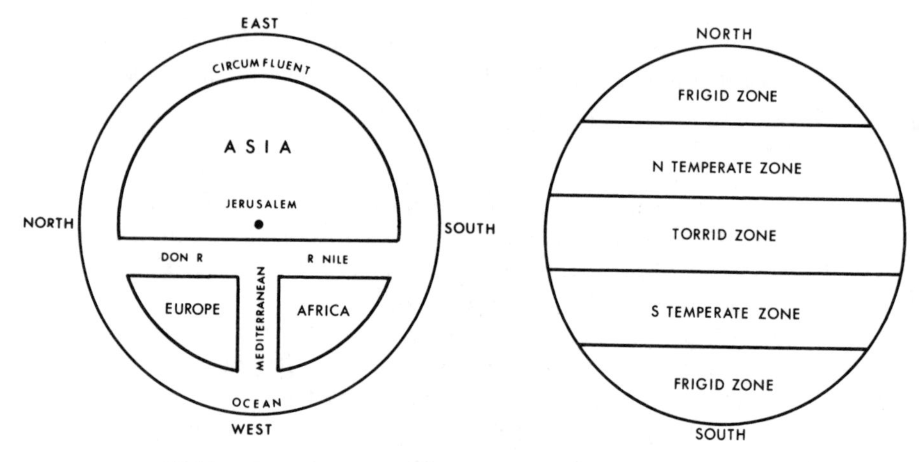

FIGS. 4.2 *and* 4.3. *Diagrams of the T-O concept* (left) *and the zonal concept* (right).

In the zonal type, the debt to Greek science is more explicit and the examples frequently have orientations other than to the east. As the ancients were concerned with the extent of the inhabited earth (*oikoumene*), which they attempted to delimit with *klimata,* so medieval scholars were interested in the area of human occupance, but for theological reasons. Both the T-O and the zonal types of circular world maps, which often accompany medieval manuscripts, are also found in the cartography of Islam; in some instances, both concepts are combined on a single map.

The ultimate expression of the circular world map of the Middle Ages is found in the Ebstorf and the Hereford maps. Both of these maps are examples from the late thirteenth century and they are similar in conception, though quite different in detail. These *mappae mundi* seem to have served as altar pieces (or perhaps were hung behind the altar) in the German monastery church of Ebstorf in the one case, and at Hereford Cathedral in England, in the other. Unfortunately, the Ebstorf map was destroyed during World War II but good color copies of this large manuscript map, which was about 13 feet in diameter, survive.[4] The Hereford map (Fig. 4.4), which is drawn on vellum, possibly a bullock's skin, is also large, measuring 5 feet 3 inches high and 4 feet 6 inches wide. It is one of the treasures of Hereford Cathedral, where it is now housed in a large case in the north aisle, and is an important survivor

[4] Bagrow-Skelton, *History of Cartography* (Cambridge, Mass.: Harvard University Press) pp. 72–73, 223. It has appeared as a jigsaw puzzle.

FIG. 4.4. *Photograph of the original manuscript Hereford world map.*

of a type of map probably found in a number of the great religious houses of Europe in the Middle Ages.

The Hereford map was formerly thought to be something of an oddity of little geographical value, but, as the result of recent research, this opinion is being revised.[5] According to an inscription on the map which requests the prayers of the viewers for the author, it was made by Richard of Haldingham and Lafford. Richard may have brought the map with him when he came from Lincoln to Hereford toward the end of the thirteenth century, though some believe that it was made in Hereford. It is a modified form of the T-O map (shown in diagrammatic form in Fig. 4.2) and has an affinity with earlier examples including those of Isidore of Seville (c.A.D. 600) and Henry of Mainz (A.D. 1100). The Hereford world map is based on classical itineraries and later sources, and can be regarded as a summary of the geographical lore, secular and sacred, of the Middle Ages. Mythical creatures and abnormal people from the fabulists are depicted, but it contains, especially in the European section, new information derived from medieval commercial journeys, pilgrimages, or Crusades. In its coloring and calligraphy, as well as in its view of the earth, the Hereford *mappa mundi* beautifully expresses the feeling of the later Middle Ages which reached its highest architectural development in the Gothic cathedral. In fact, there are close relationships between the symbolization used in these two forms. For example, the representation of Christ in Judgement above the circumfluent ocean on the Hereford world map is reminiscent of the tympani in contemporary ecclesiastical architecture (Fig. 4.5). The celestial world with its perfection is set apart and above the imperfect terrestrial earth. Although some savants in the Middle Ages, such as the Venerable Bede and Roger Bacon, investigated physical phenomena, at times, there were proscriptions against this, including the measuring of the depths of the sea.

It has been debated whether such works as the Hereford *mappa mundi* were intended to aid travelers or as inspirational pictures like the stained glass windows of the cathedrals. No doubt they served both functions and some pilgrims who looked at them probably suggested changes based on their own travels. However, there were medieval maps more specifically designed for the assistance of travelers. Thus, at the scriptorium of St. Albans, the monk Matthew Paris made not only a *mappa mundi* and a map of Great Britain, but also a strip map showing pilgrimage routes within England, and another showing routes from

[5] Gerald R. Crone, "New Light on the Hereford Map," *Geographical Journal,* vol. 131 (1965), 447–62; by the same author, *The World Map of Richard of Haldingham in Hereford Cathedral* (London: Royal Geographical Society, 1954). The latter contains a large monochrome reproduction of the Hereford map in nine sheets.

FIG. 4.5. *Detail of the upper part of the Hereford world map.*

London to southern Italy. Such strip maps resemble the Peutinger Table in that routes are indicated by straight lines from place to place, with no special attention paid to orientation.

While these developments were taking place in the religious houses in medieval Europe, map-making was progressing elsewhere. We have already discussed contemporaneous mapping activities in the Far East, but scholars from the Middle East also contributed to cartography in this period.[6] After the fall of Babylon, science in the arid lands of southwest Asia appears to have been strongly influenced by India. This is particularly true before the reformation of the Arabic alphabet and the translation of Ptolemy's works into Arabic in the ninth century A.D. In the next century, Ptolemy's maps became available to the Arabs. Following this, two distinct developments affecting cartography unfolded in Islam: (*1*) the determination of the latitude and longitude of places on the earth as part of an increasing emphasis on astronomy; and (*2*) geographical descriptions arising from the extensive land and sea journeys undertaken for conquest, administration, or trade. Route maps were prepared to facilitate these latter Moslem undertakings.

Although at first Arab scholars seem to have accepted Ptolemy's astronomical works, in time they were criticized and improved. The length of the Mediterranean Sea, given by Ptolemy as 62 degrees, was reduced to 52 degrees by Al Khwarizimi (ninth century A.D.) and further reduced to its correct figure of 42 degrees, through the observations of Al Zargālī (twelfth century A.D.). Among the Arab travelers who supplied accounts of their journeys, Ibn Baṭṭūṭa (A.D. 858–929) and Al Idrīsī (A.D. 1099–1166) are outstanding examples.

It has been said, perhaps unfairly in view of the fact that much of the cartographic record must be lost, that the Arabs were better astronomers and geographers than cartographers. To illustrate Islamic cartography, we will use the work of Idrīsī who, after undertaking extensive travels himself, was invited to Sicily by its enlightened Norman King, Roger II. In Sicily, Idrīsī engaged in geographical writing and in the compilation of maps. He made a circular world map (Fig. 4.6) with curved parallels which, in a number of respects, is superior to contemporaneous European maps of the same genre. His most important work, however, is a large rectangular world map, in seventy sheets, known as the *Tabula Rogeriana.* Figure 4.7 is a reproduction of a page of Idrīsī's atlas showing the Aegean Islands, while Figure 4.8 is a redrawing of a large section of Idrīsī's world chart of A.D. 1154, the earlier and generally

[6] Carl Schoy, "The Geography of the Moslems of the Middle Ages," *Geographical Review,* vol. 14 (1924), 257–69. In Arabic works, cartography is often discussed with geography (*djughrafiya*), with no distinction made between the two studies.

FIGS. 4.6 *and* 4.7, opposite. 6 (top), *Arabic zonal world map, with south orientation, by Idrīsī.* 7 (bottom), *Peloponnisos, Kikladhes, and Kríti (Crete): A small section of Idrīsī's world map in atlas form, with south orientation.*

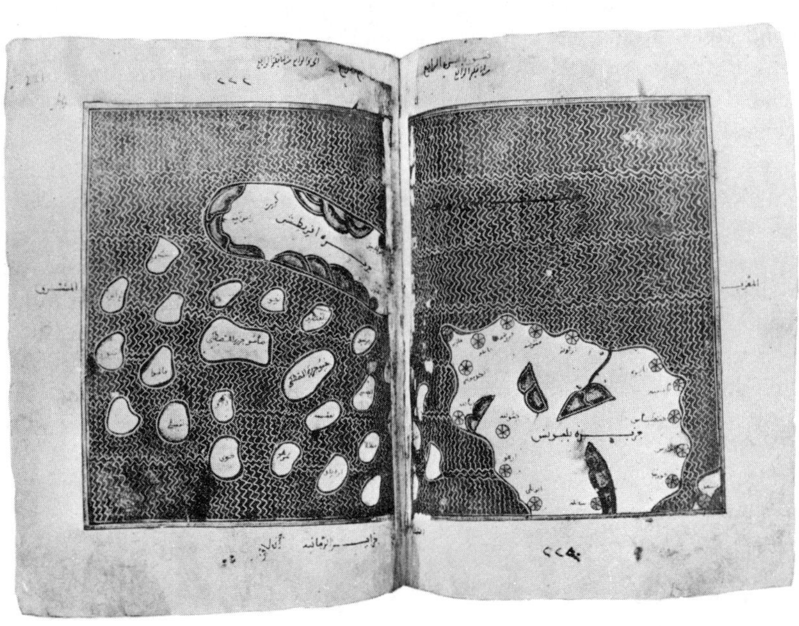
المشرق
المغرب

better of two such maps that have come down to us through the work of copyists. It is instructive to compare Idrīsī's south-oriented map with the Hereford map (Fig. 4.4) of approximately the same time. Obviously, the *Tabula Rogeriana* is less stylized, and it incorporates new information supplied from the travels of Idrīsī and many others. Some of the cartographic work executed at the Norman Court in Sicily, including a map engraved on a silver tablet, has perished; but the surviving record demonstrates the originality of Idrīsī's contributions which continued to be important in the Arab world for centuries after his death. The extent of the influence of Ptolemy's work on Idrīsī has been speculated upon, but takes nothing away from latter's accomplishments.

We mentioned the magnetized needle in the discussion of Chinese cartography and speculated on its use in mapping in that area and on the transmission of this useful instrument to Europe. In the later Middle Ages, there was great interest in Europe in the properties of the lodestone, and experiments involving magnetism were performed by Roger Bacon and others. Various attempts were made to satisfactorily mount the needle and apparently this was accomplished in Amalfi, Italy at the end of the thirteenth century. In the Mediterranean, the compass—consisting of a box containing a pivoted, magnetic needle mounted over a card on which sixteen, and later thirty-two, directions were painted—came into use among sailors. Because the directions were named for winds, following the practice of antiquity, such geometrical constructions are known as wind roses.[7] With the help of the compass, great progress in mapping and in navigation was possible, and a new cartographic form presumably related to this development appeared in c. 1300—the portolan chart, (Fig. 4.9).

The origin of the portolan chart is obscure, but it seems to be an obvious extension of the descriptions found in pilot books (*portolani*). That the earliest surviving examples are well developed suggests that even earlier portolan charts may be lost. They are typically drawn on a single sheepskin and oriented according to the magnetic north. Since their purpose was to aid navigators, shorelines are emphasized, and little geographical information appears on the land. Characteristically, portolans show the Mediterranean and Black Sea coasts with remarkable accuracy, though with curiously stylized symbols in detail. Place names are

[7] Silvanus P. Thompson, "The Rose of the Winds: The Origin and Development of the Compass-Card," *British Academy Proceedings, 1913–1914* (London, 1919), pp. 179–209; and Norman J. W. Thrower, "The Art and Science of Navigation in Relation to Geographical Exploration," in *The Pacific Basin*, Herman Friis, ed. (New York: American Geographical Society, 1967), chap. 2, pp. 18–39, 339–43.

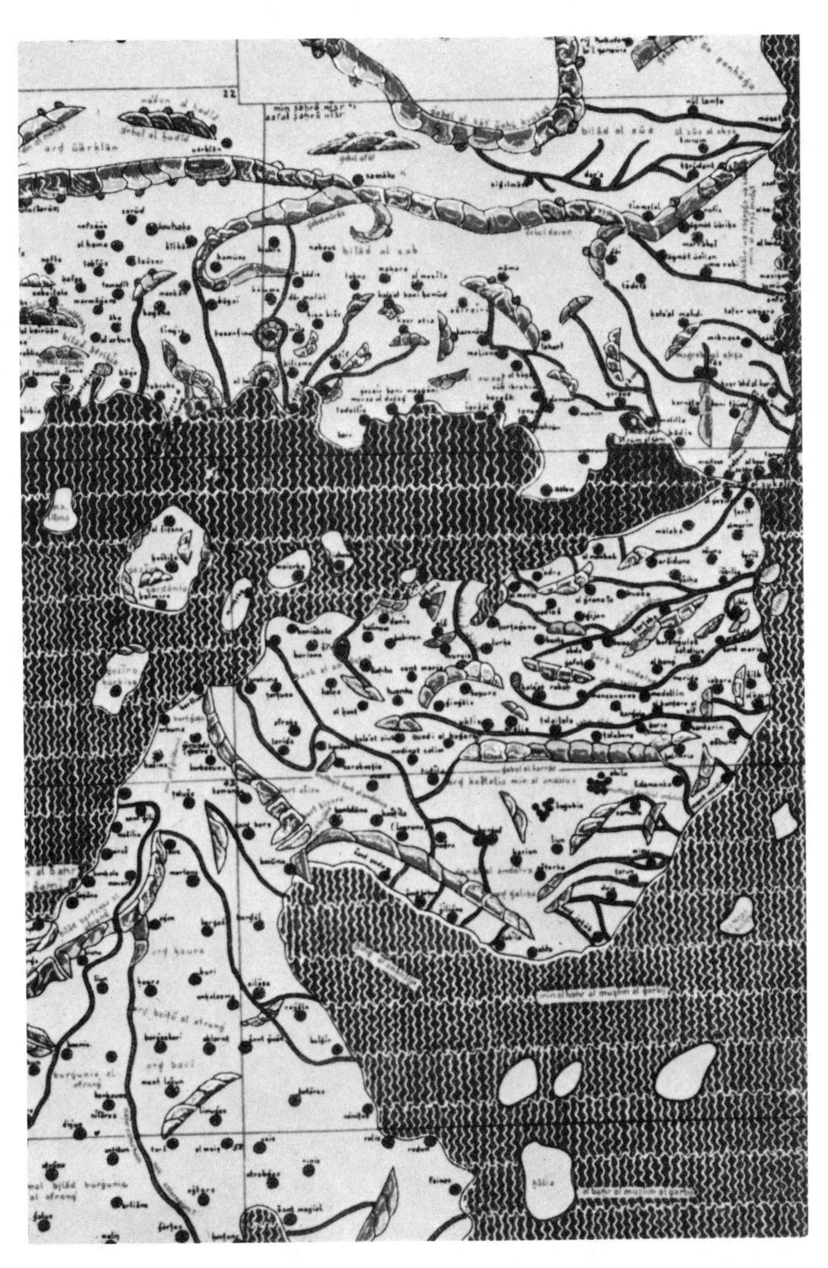

FIG. 4.8. *Portion of Idrīsī's world map, with south orientation.*

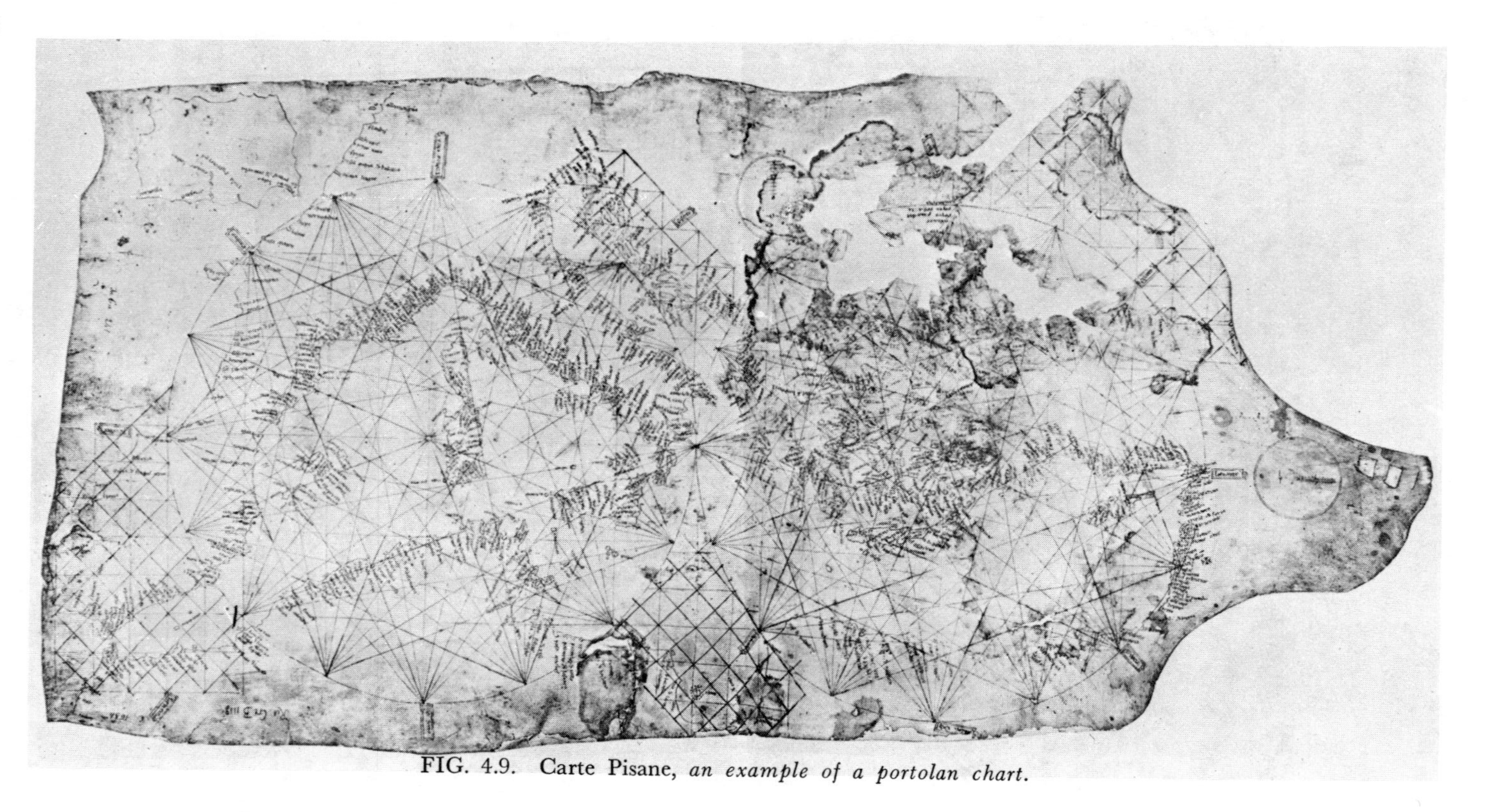

FIG. 4.9. Carte Pisane, *an example of a portolan chart.*

lettered perpendicular to the shore. Striking features of these maps are the representations of wind or compass roses and lines emanating from them which crisscross the charts. These lines were apparently used to assist the navigator to plot compass bearings, with the aid of a parallel ruler, before true or astronomical bearings came into general use. Although basically practical sea charts, the portolans, especially late ones, are richly illuminated with gold, silver, and various colors on the coats of arms, flags, and miniature paintings of cities with which they are decorated. Almost a conventional color scheme is used for the lines extending from the compass roses: black for the eight principal winds; green for the eight half winds; and red for the sixteen quarter winds.[8]

To illustrate this cartographic form, we have selected one of the many surviving portolans, the famous *Carte Pisane* or Pisan Chart (Fig.

8 Adolf E. Nordenskiöld, *Periplus, An Essay on the Early History of Charts and Sailing Directions* (Stockholm: P. A. Norstedt & Soner, 1897). Although some of Nordenskiöld's ideas on this subject have been discredited, this work by one of the founders of the study of historical cartography should be consulted. See also, Edward L. Stevenson, *Portolan Charts, Their Origin and Characteristics* (New York: The Knickerbocker Press, 1911). The last phases of portolan chart-

FIG. 4.10. *The shore of southern Italy and Sicily (represented by the solid line), redrawn from the* Carte Pisane; *the dotted line is the same shore from a modern map.*

4.9 and detail 4.10). This oldest surviving portolan chart (c. 1300), as its name suggests, is of Italian origin. Like most portolans from this area, it shows only the Mediterranean and Black Sea region. However, the maritime cities of Italy were not the only centers of chart-making at this time. Portolan charts embracing the Mediterranean, northern Europe and other parts of the world were made by Catalan (including Jewish) cartographers of Majorca and Barcelona, in the service of the Kings of Aragon. In time, coastlines were extended and information from Arab and other sources was added so that portolan-type world maps eventually developed. The famous and very rich Catalan Atlas (A.D. 1375) of Abraham Cresques is one example of this process.

---

making have recently been discussed by Thomas R. Smith, "Nicholas Comberford and the 'Thames School'—Sea Chart Makers of Seventeenth Century London" (unpublished paper read at the Third International Conference in the History of Cartography, Brussels, 1969).

The Rediscovery of Ptolemy and Cartography in Renaissance Europe

5

The cartographic work of Claudius Ptolemy was discussed in connection with contributions from antiquity, and his influence on Islamic science considered in the preceding chapter. When the Turks expanded westward and reached Byzantium (Constantinople) in the fourteenth century, refugees fled from that city, taking with them various treasures—including the Greek texts of Ptolemy's *Geographia*. These manuscripts reached Italy, and by 1410 were translated into Latin in Florence (Firenze) where geography and mathematics, as well as art, were important studies.[1] Thus manuscript copies of the Latin translation of the *Geographia*, first without maps and later with maps, became available.

The influence of Ptolemy's work on Western cartography would be hard to exaggerate; it is only necessary to compare the best Medieval *mappae mundi* with the Ptolemaic maps to appreciate the general superiority of the latter. This did not prevent European scholar-cartographers from soon improving upon certain areas and features of Ptolemy's maps, as, for example, Claudius Clavus in his map of Scandinavia of 1427, and Fra Mauro in his world map of 1457. Ptolemy's *Geographia* appeared in the Latin West at a time when important developments were taking place which greatly affected the progress of cartography. The most important of these were the invention of printing in Europe, and the expansion by Europeans to overseas areas.

The importance of printing to cartography lies not only in the reduction of the cost of maps (actually

1 Thomas Goldstein, "Geography in Fifteenth-Century Florence," in *Merchants or Scholars,* John Parker, ed. (Minneapolis: The University of Minnesota Press, 1965), chap. 1, pp. 11–32; and Leo Bagrow, "The Origin of Ptolemy's Geographia," *Geografiska Annaler,* vol. 27 (1945), 318–87.

in some circumstances prints might be more expensive than hand-drawn examples), but more especially in the ability to produce copies that are essentially alike. We have already considered the prior invention of printing in China and its application to cartography as early as A.D. 1155. The first European printed maps date from the last three decades of the fifteenth century. These incunabula include simple woodcut reproductions of various T-O and zonal world maps, engraved copies of Ptolemy's world, and sectional maps, from both Italy and Germany. To illustrate this development, we reproduce the map of the world according to Ptolemy from the Latin edition of the *Geographia* published in Ulm in 1482 (Fig. 5.1). This is neither the earliest map printed in Europe (this distinction belongs to a simple woodcut of St. Isidore's T-O map, printed in Augsburg in 1472), nor the earliest printed Ptolemaic map, which was published in Bologna in 1477. However, perhaps better than any other map, the Ulm example typifies Ptolemaic cartography. It is drawn on a modified spherical projection—an ingenious solution to the problem of representing the allside curving figure of the globe (or, at least, a large part of the earth) on a flat surface. This is the most complicated of the projections devised by Ptolemy, who also used a simple conic form for world and sectional maps (Fig. 5.2). At the margins of the Ulm map (Fig. 5.1), twelve wind blowers indicate direction. This is an elaboration of the concept of four wind directions of antiquity that was increased to eight (as used on early compasses to divide the circle of the horizon) and later, to sixteen and, eventually, thirty-two. Other features of the Ulm map include rivers, lakes, and mountain ranges.

Without going into great detail concerning land and water relationships on Ptolemy's world map (Fig. 5.1), we should note that the then known inhabited earth (Old World) extends about 180 degrees, halfway across the sphere in the northern mid-latitudes, whereas, in reality, it covers only about three-eighths of this distance. Other features include an enclosed Indian Ocean with no apparent sea route around southern Africa to India, a comparatively good representation of the Malay Peninsula, and a truncated India with an exaggerated Ceylon (Taprobane). The delineation of the Mediterranean is only fair in comparison to existing portolan charts. This region, and northern Europe, which is poorly represented in Ptolemy's maps, were soon criticized and improved. As the Europeans extended their sphere of influence overseas, the conception of the Ptolemaic world was shortly to be radically altered.

FIG. 5.1. (opposite) *World map from the Latin edition of Ptolemy's* Geographia, *published in Ulm, 1482.*

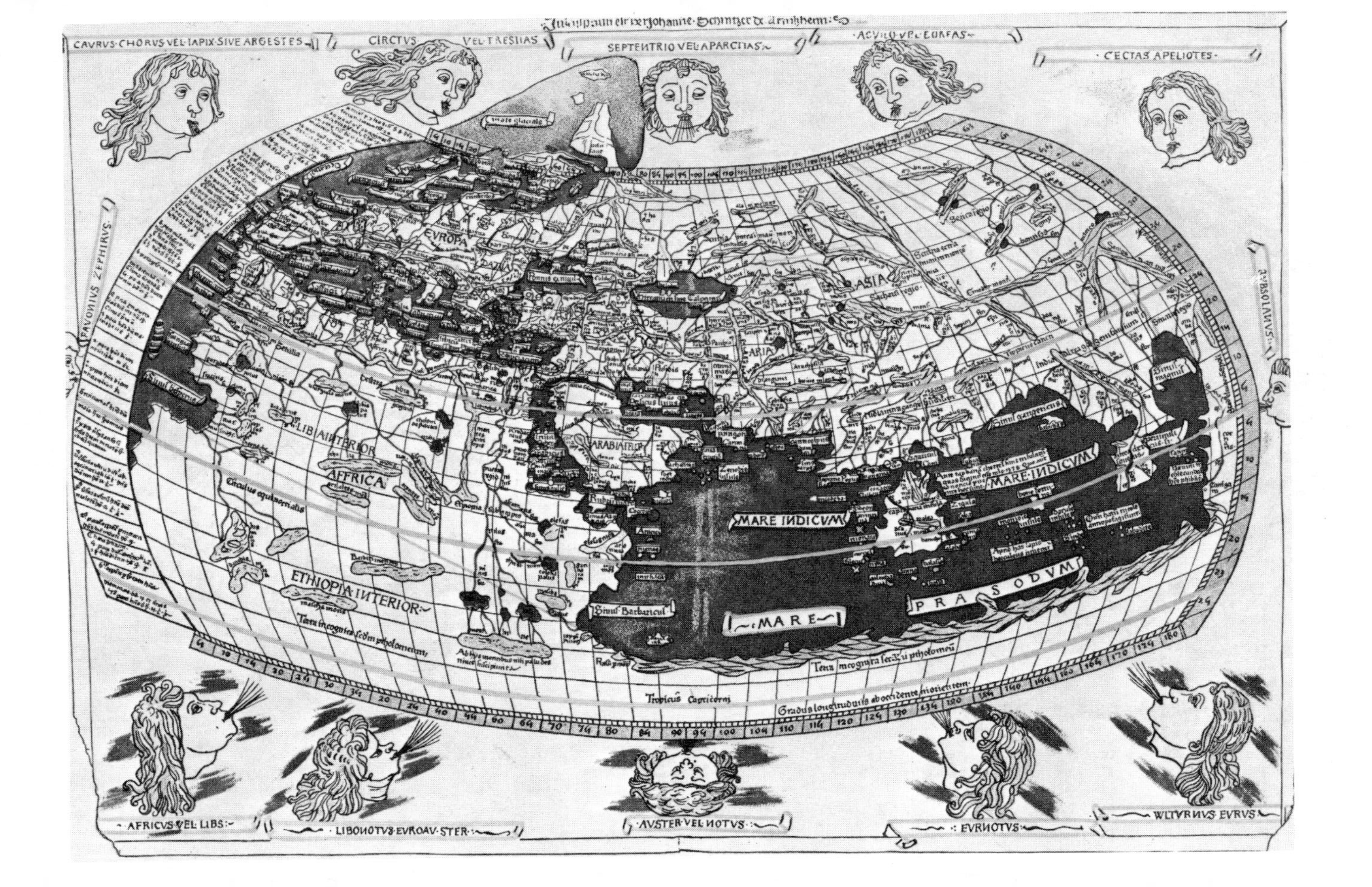

Insculptum est per Johanne Schnitzer de Armßheim
CAURUS CHORUS VEL IAPIX SIVE ARGESTES
CIRCIUS VEL TRESCIAS
SEPTENTRIO VEL APARCTIAS
AQUILO VEL BOREAS
CECTAS APELIOTES
FAVONIUS ZEPHIRUS
SUBSOLANUS
EUROPA
ASIA
AFFRICA
LIBIA INTERIOR
ARABIA FELIX
ETHIOPIA INTERIOR
MARE INDICUM
PRASODUM
MARE
Sinus Barbaricus
Circulus equinoctialis
Tropicus Capricorni
Terra incognita secundum ptholomeum
Gradus longitudinis ab occidente in orientem
AFRICUS VEL LIBS
LIBONOTUS EUROAUSTER
AUSTER VEL NOTUS
EURNOTUS
VULTURNUS EURUS

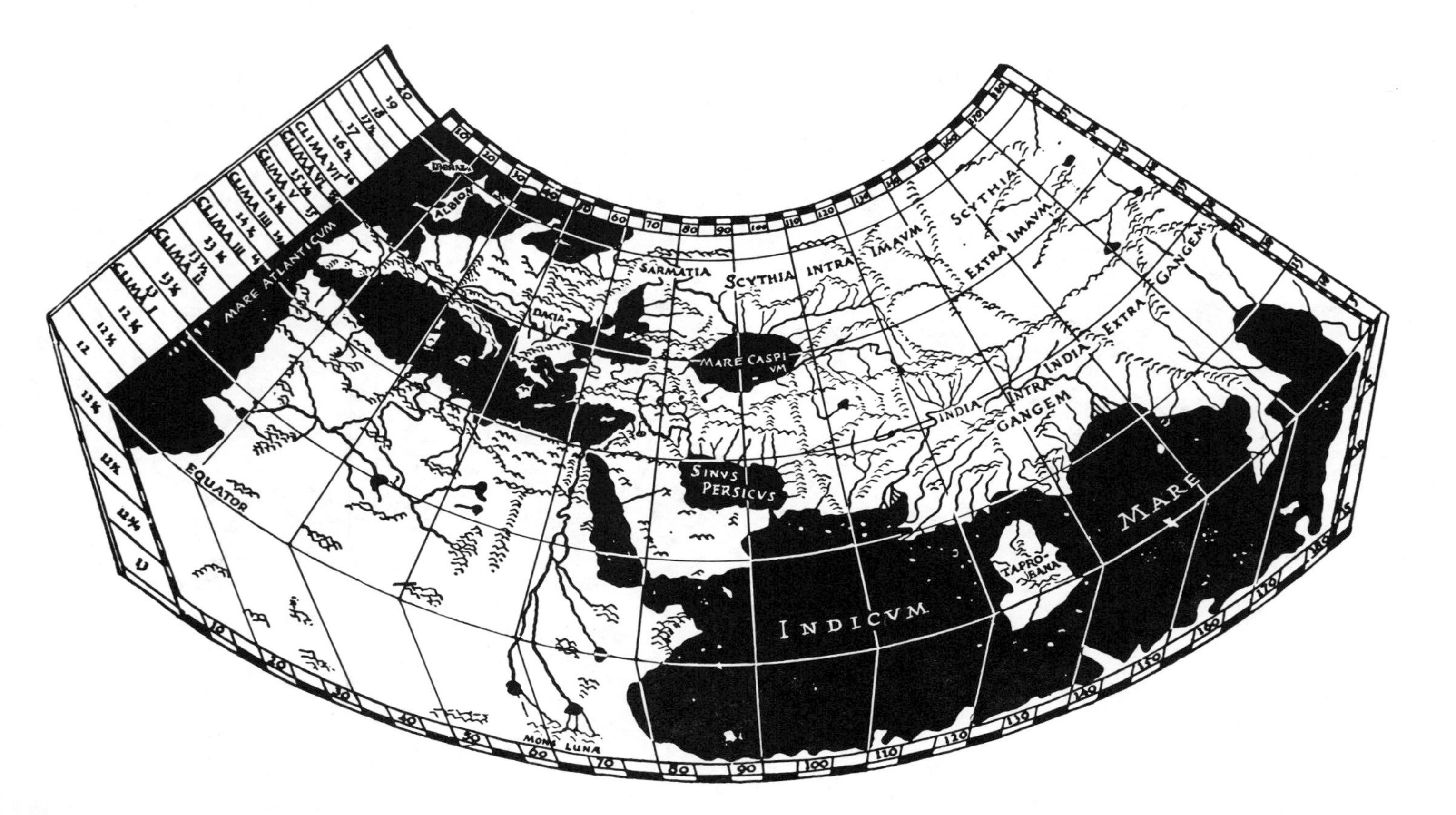
MARE INDICUM
TAPRO-BANA
INDIA EXTRA GANGEM
INDIA INTRA GANGEM
SCYTHIA EXTRA IMAVM
SCYTHIA INTRA IMAVM
SARMATIA
MARE CASPI VM
SINVS PERSICVS
DACIA
ALBION
MARE ATLANTICUM
EQUATOR
MONS LUNA
CLIMA I
CLIMA II
CLIMA III
CLIMA IIII
CLIMA V
CLIMA VI
CLIMA VII

The rediscovery of Ptolemy's work in the West coincided with the beginning of the age of the great European overseas discoveries—the two events are, in fact, related. This exploring activity was formally initiated by Prince Henry of Portugal (b. 1394), later called The Navigator.[2] After 1419 when he settled in the Algarve, the most southwesterly province of Europe, Henry enlisted the aid of all who could assist him in his work—sailors, shipbuilders, instrument makers, and cartographers. These last included a "master of sea charts," presumably Master Jacome, son of Abraham Cresques from Majorca who brought the Catalonian portolan tradition to Portugal. In addition, one of Henry's older brothers, Prince Pedro, visited Italy, including Florence in 1428, for the purpose of collecting maps and new geographical information.[3]

Meanwhile, under Henry's auspices, various expeditions rediscovered the Azores, which appeared on maps he had received, and probed southward along the west coast of Africa. The Cape Verde Islands were reached and Islam outflanked before Prince Henry's death in 1460. By the end of the fifteenth century, America had been discovered and the sea route to India found. During the next 300 years, most of the coasts of the world were discovered by the explorers of Portugal, Spain, Italy, Holland, France, and Britain. Through exploitation of these discoveries, Europe grew from a poor peninsula of Eurasia in the Middle Ages to the most influential area of the world in the seventeenth, eighteenth, and nineteenth centuries.[4] Exploring activities produced a great body of information that, sooner or later, was added to maps. Much of the history

[2] Charles R. Beazley and Edgar Prestage, eds., "G. Eannes de Azurara: The Chronicle of the Discovery and Conquest of Guinea," *Hakluyt Society Publications,* series 1, vols. 95, 100 (London: printed for the Hakluyt Society, 1896–99); Norman J. W. Thrower, "Prince Henry the Navigator," *Navigation,* vol. 7, nos. 2, 3 (1960), 117–26. This article was one of several printed in various journals commemorating the fifth centenary of the death of Prince Henry, which was marked by special events in Portugal and elsewhere.

[3] Goldstein, "Geography in Florence," 17–18. One of these maps is believed by some to be a copy of Marino Sanudo's *Mappa Mundi,* dated c. 1320, on which the Azores are shown.

[4] In recent years, there has been some reaction against the claims of European discovery of the world. A short time ago, some Zambians indicated that their ancestors "discovered" David Livingstone in the 1850s. Similarly, several Ethiopians solved the "Rhine Problem" and handed out trinkets to the natives during their ascent of that river. An attempt to put geographical discoveries in proper perspective appears in Herman Friis, ed., *The Pacific Basin* (New York: American Geographical Society, 1967), where the exploring activities of the Pacific Islanders, Chinese, Japanese, and other non-Europeans are considered, along with those of the Portuguese, Spanish, Dutch, etc.

FIG. 5.2. (opposite) *Redrawing of Ptolemy's world map on his simple conic projection.*

of cartography has been concerned with the "unrolling" of the world map and there is a vast literature dealing with European geographical exploration and its cartographic representation.[5] The two subjects are closely interrelated because a place is not really discovered until it has been mapped so that it can be reached again.

A cartographic work which has recently aroused much controversy is the Vinland Map, claimed to have been drawn "in the 1440's." Supposedly for the first time it charted Vinland (part of the east coast of North America) discovered by Bjarni Herjolfsson (C.A.D. 985) and Leif Eriksson (C.A.D. 1000) as an inscription on the map testifies.[6] Many scholars now affirm the Vinland Map to be a twentieth century compilation of voyages described in early sagas.

Important as these discoveries might be, we will not deal at great length with this matter here, but only review some of the highlights of European discovery in order to appreciate the cartography of the Renaissance. This will be done with particular reference to just three world maps which in a sense summarize the exploring activities of the Europeans during the last quarter of the fifteenth century and the first three-quarters of the sixteenth century—100 years during which the European perception of the world land and water relationships changed more than it did in any comparable period. The few maps by explorers from this period that have survived are, characteristically, sketches of small areas, and it is more instructive for our purpose to illustrate growing knowledge of the world with maps by well-known cartographers who incorporated the findings of the discoverers in their work.[7] Understandably, these maps better illustrate developing cartographic methodology than those of the explorers themselves.

The first of the three maps (Fig. 5.3) is a reproduction of the gores of the manuscript *erdapfel* (world globe) by Martin Behaim of

[5] In addition to a large number of general works on exploration, the publications of the Hakluyt Society have focused "on original narratives of important voyages, travels, expeditions and other geographical records." Since its founding in 1846, the Society has published approximately 250 titles, many of which contain maps. An organization with similar objectives, The Society for the History of Discoveries, was founded in the United States in 1960; the first volume of its journal, *Terrae Incognitae,* appeared in 1969. For a popular and graphic account of the progress of European discoveries, see Leonard Outhwaite, *Unrolling the Map* (New York: John Day, 1939).

[6] R. A. Skelton, Thomas E. Marston, and George D. Painter, *The Vinland Map and the Tartar Relation* (New Haven: Yale University Press, 1965).

[7] R. A. Skelton, *Explorers' Maps* (London: Routledge and Kegan Paul, 1958); by the same author, "Map Compilation, Production, and Research in Relation to Geographical Exploration," chap. 3 in *The Pacific Basin,* pp. 40–56, 344–45. See also Armando Cortesão and Avelino Teixeira de Mota, *Portugaliae monumenta cartographica* (Lisbon: Comissão Executiva das Comoracões do V Centenario da Morte do Infante D. Henrique, 1960).

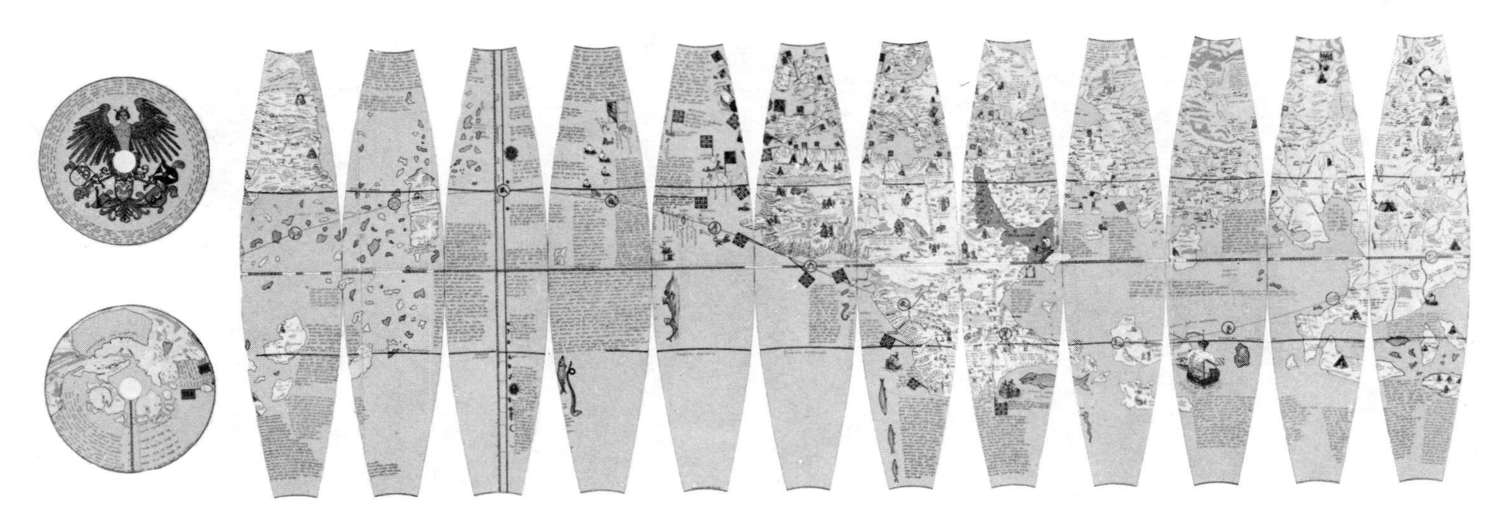

FIG. 5.3. *Martin Behaim's* erdapfel: *gores of the globe and polar caps.*

Nürnberg who spent some time in Portugal and who apparently had been engaged in explorations along the African coast.[8] The globe, like the map of Henricus Martellus Germanus of approximately the same date, shows the world just prior to the discovery of America by Columbus. The debt to Ptolemy is obvious and acknowledged, but Behaim includes new information on eastern Asia, resulting, particularly, from the descriptions of those areas by Marco Polo. The peninsular character of southern Africa is the most striking difference between the work of Behaim and Ptolemy; the then new discoveries of the Portuguese, notably Diaz, were responsible for this improvement. India and Ceylon were remarkably unchanged in comparison with the earlier delineations of these areas.

Cartographically, Behaim's globe, which was constructed and painted by superb craftsmen, is of considerable interest. A globe is, of course, the most accurate means of representing the earth, but it is not the most useful device for all purposes because of the difficulty of measurement and of ascertaining geographical relationships on a sphere. Even when spread out as a series of gores so that the whole world can be seen at one glance, the interruptions (caused in this case by the cuts along poleward converging meridians) create map-reading problems. Behaim's globe has a diameter of 20 inches and is divided into twelve gores of 30 degrees each; the entire 360 short degrees are marked where the gores are hinged at the equator. The Tropics, the Arctic and Antarctic circles are shown, but only the 80 degree meridian is drawn from pole to pole. Eurasia, in the northern hemisphere's mid-latitudes, covers approximately three-quarters of the globe. As a result, the distance between Europe and Asia across the Atlantic is comparatively short. It was perhaps a prototype of Behaim's globe, but not this work itself (which he could not have seen), that encouraged Columbus, who was a chartmaker as well as an explorer, to venture westward in the hope of finding the coast of Asia.

Behaim's *erdapfel* is richly colored—water bodies are blue, except for the Red Sea which is vermillion on this and other maps of the period; land is generally buff or ochre with gray stylized, side-viewed mountains, and indications of forests are green. Ships, sea creatures, zodiacal signs, and flags (portolan style) are rendered in color. The work was left behind, according to the author, "for the honor and enjoyment of the commonalty of Nürnberg."

The second illustration of the changing view of the earth and of cartographic progress as well, is a world map by Martin Waldseemüller, dated 1507 (Fig. 5.4). This is a print of one of a number of woodcut maps by this famous cartographer who worked at Saint-Dié in the

[8] E. G. Ravenstein, *Martin Behaim: His Life and His Globe* (London: George Philip & Son, 1908).

Rhineland. The map frame is suggestive of Ptolemy's spherical projection although it is actually a special case of the cordiform (heart-shaped) projection, which, in uninterrupted and interrupted form, became popular in the Renaissance.[9] The maps of Waldseemüller are of particular interest because his were apparently the first on which the name America appears; in fact, it was Waldseemüller who suggested the name America for the New World in honor of Amerigo Vespucci, who explored and named the Venezuelan coast in 1499. The Americas, as represented by Waldseemüller, are of very limited longitudinal extent because they are fitted into the Ptolemaic longitudinal framework.

At this time (1507), only the east coasts of the New World were known through the explorations of Columbus, Cabot, Cabral, and Vespucci. These discoveries are shown on a map drawn in 1513 by the Turkish admiral, Piri Re'is, and thus presumably soon became known in the Moslem world. The sea route to India, by way of the Cape of Good Hope (which had been discovered by Vasco da Gama in 1497 with the help of an Arab pilot of Malindi on the east African coast), was well known to Europeans by 1507.

In 1508, Vespucci became the first Pilot-major or supervisor of maps and charts of the *Casa de la Contratación de las Indias* in Seville. The responsibilities of this institution included: the entering of new information on a master map (the *Padrón General* or official record of discoveries); the supervision of charts and instruments carried by seamen; and the examination of pilots. Although the *Padrón* is lost, it is believed to be approximated by the world chart of Diego Ribero (1529). The Spanish *Casa* was founded at the beginning of the sixteenth century, but since the end of the previous century a hydrographic office with similar objectives, the *Casa da Guiné* (later *Casa da India*) had existed in Portugal. In time, through conservatism, these bodies tended to inhibit rather than aid cartographical and navigational progress.

The hallmark of the Renaissance was the universal genius, and we find that a number of men who are primarily thought of in other connections were engaged in cartography—e.g. Leonardo Da Vinci. Those more specifically associated with mapping include Giovanni Contarini, Sebastian Münster, Peter Apian, and Gemma Frisius. Frisius, an astronomer and mathematician as well as a cartographer, was the teacher of Gerhardus Mercator. Mercator (Kremer), the author of the next map to be considered, was born in Flanders in 1512. While at the University of Louvain, Mercator became a student of Gemma for whom he engraved

[9] George Kish, "The Cosmographic Heart: Cordiform Maps of the 16th Century," *Imago Mundi,* vol. 19 (1965), 13–21, is an interesting article on this type of projection of which a number of forms were devised by Renaissance cartographers.

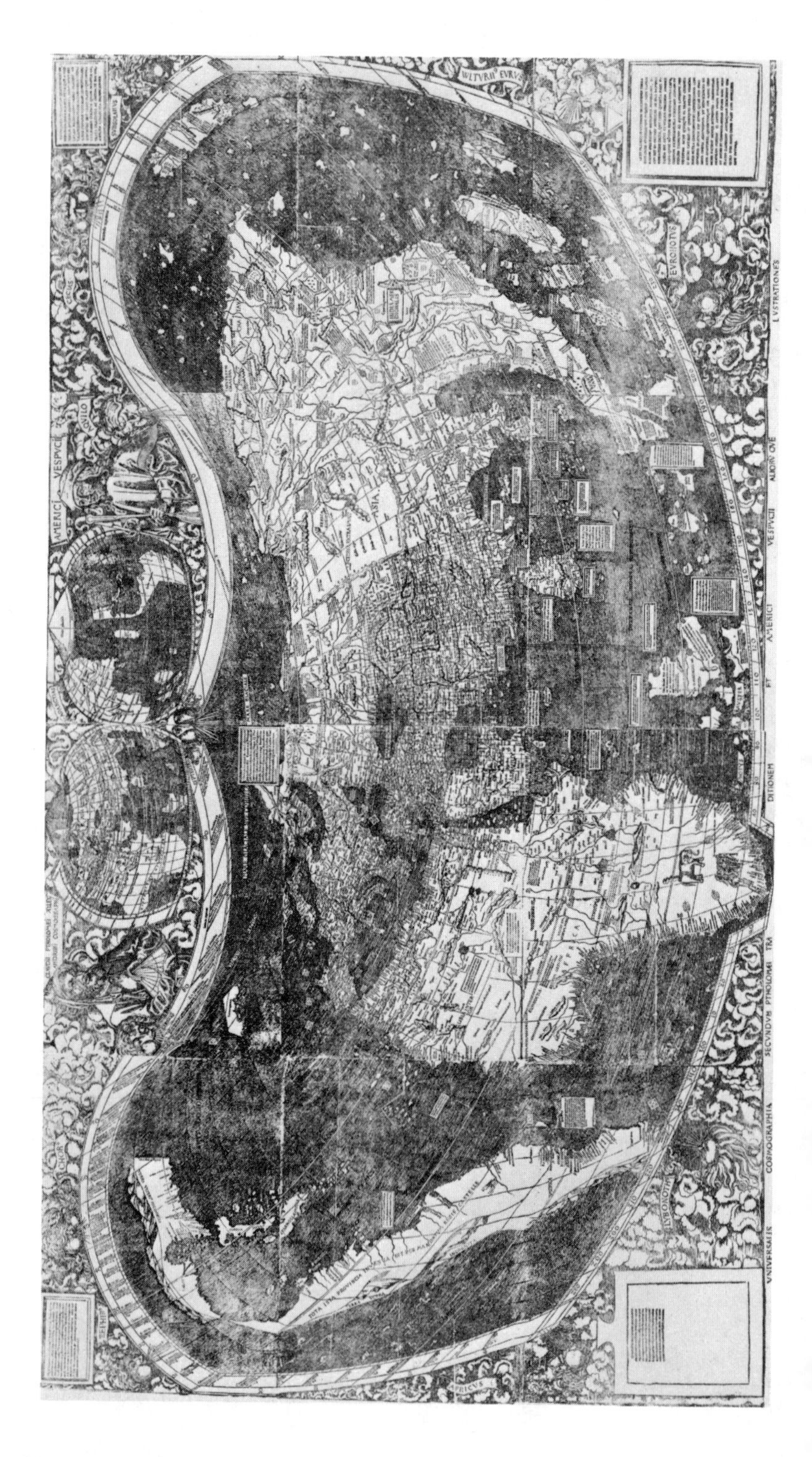

a globe (c. 1536). He went on to publish a world map on a double cordiform projection in 1538, and also engaged in land surveying, producing highly accurate maps of Europe in 1541 and 1554, the latter being on a conical projection with two standard parallels. Mercator was an expert engraver and introduced italic lettering to northern Europe.[10] By utilizing the best available itineraries and charts as source materials, Mercator reduced the map length of the Mediterranean from Ptolemy's figure of 62 to 52 degrees which, though still too great by 10 degrees, nevertheless represented a great cartographic improvement. Since antiquity, latitude had been measured with considerable accuracy, but longitude, especially on shipboard, was difficult to determine before the invention of reliable, portable time pieces in the second half of the eighteenth century.[11]

All of the previous accomplishments of Mercator, who was established in Duisburg in the Rhineland, were eclipsed by the publication (in 1569) of a great world map on the projection that bears his name (Fig. 5.5). This projection is a superb example of the positive value of a map over a globe for a specific purpose. Like several other projections, the Mercator is conformal (shapes around a point are correct), but it also has a unique property—straight lines are rhumb lines or loxodromes (lines of constant compass bearing). This quality, which makes the projection of great value to the navigator, is accomplished by increasing the spacing of the parallels by specified amounts from equator to poles. Apparently Mercator derived his projection empirically and it remained for the English mathematician, Edward Wright, to provide an analysis of its properties, which he published in *Certaine Errors in Navigation* (1599). An English world map of this same date on the Mercator projection, believed to be by Wright, was bound with Richard Hakluyt's *Voyages.* It shows the results of Drake's travels and is referred to by Shakespeare in the *Twelfth Night* (Act III, Scene 2): "He does smile his face into more lynes then are in the new Mappe with the augmentation of the Indies."

10 F. Van Ortroy, "Bibliographie Sommaire de L'Oeuvre Mercatorienne," *Revue des Bibliotheques,* vol. 24 (1914), 113–48; A. S. Osley, "Mercator: A Monograph on the Lettering on Maps, etc., in the XVI Century Netherlands," (New York: Watson Guptill Publications, 1969).

11 Norman J. W. Thrower, "The Discovery of the Longitude," *Navigation,* vol. 5, no. 8 (1957–58), 374–81.

FIG. 5.4. (opposite) *Martin Waldseemüller's world map of 1507, showing the Americas with a very limited longitudinal extent.*

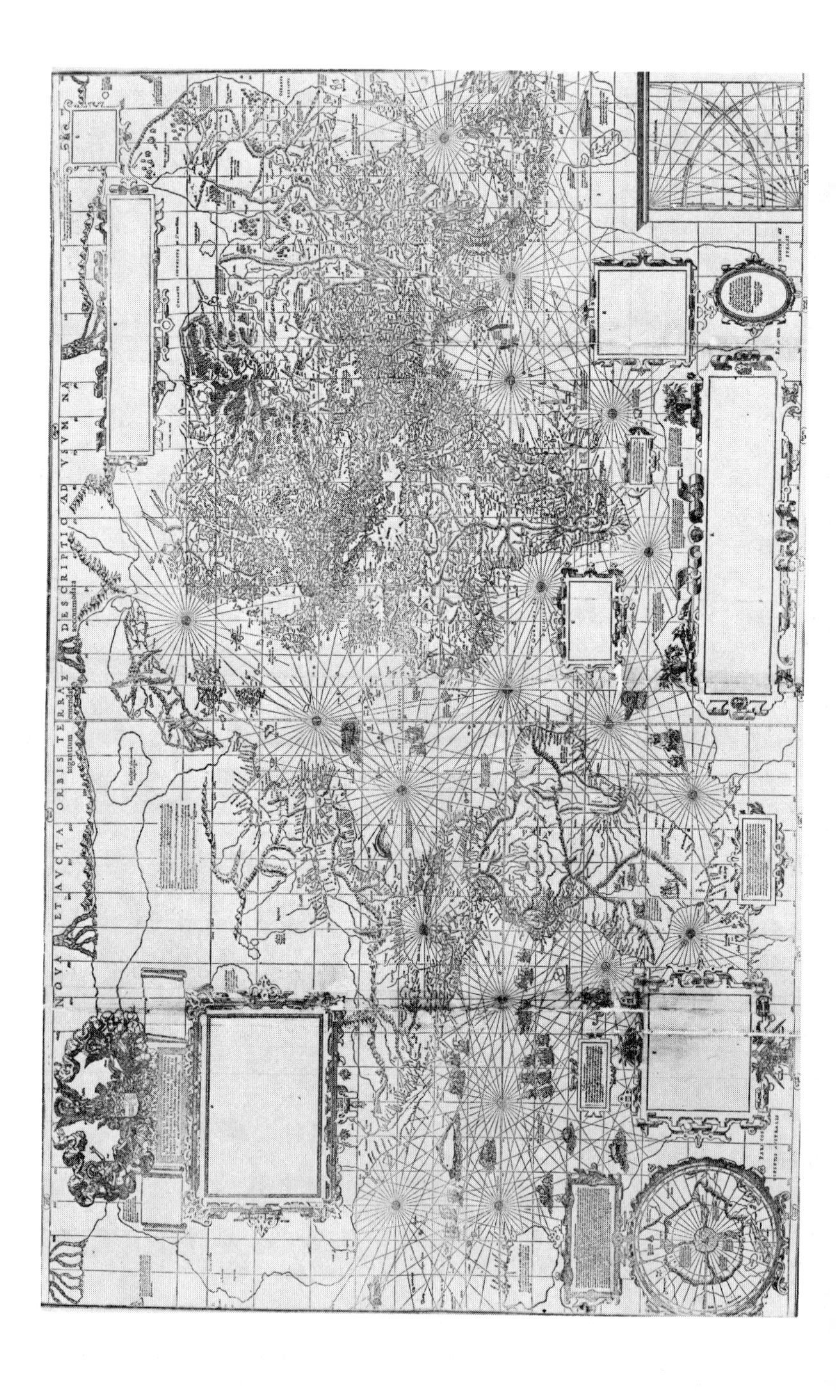
NOVA ET AVCTA ORBIS TERRAE DESCRIPTIO AD VSVM NA

There are indications that, before Mercator, portolan and Iberian plane chart-makers were struggling toward a solution of representing loxodromes. The Nürnberg instrument-maker, Erhard Etzlaub, had engraved maps (1511–13) with latitude gradations in "increasing degrees" along the edges.[12] However, Mercator was the first cartographer to produce a true navigational chart with graticules on which a compass line intersects each meridian at a constant, given angle. Although the advantages of the Mercator chart were explained by Wright and others, it appears that sailors were slow to adopt it and the plane chart remained in use for centuries.

We have been concerned with the projection used by Mercator in his map of 1569; we should now briefly consider the geographical content of that chart. Between the publication of Waldseemüller's map of 1507 and Mercator's map of some 60 years later, great progress had been made in the European exploration of the world's coastlands. In 1513, Balboa sighted the Great South Sea from a peak in Darien. Seven years later, Magellan sailed into the Pacific to be followed by a dozen major Iberian expeditions before the appearance of Mercator's map in 1569. This chart reflects much of that progress in the delineation of the west coasts of South and Middle America. Although a decorative cartouche blocks much of North America, the great longitudinal extent of this continent is suggested. Baja California is firmly attached to the continent though it was later represented as an island.[13] Southeast Asia is more accurately represented on Mercator's map than on Waldseemüller's, as the result of more than 50 years of Portuguese exploration in that area. There is also an outline of a great continent in the South Seas, *Terra Australis,* recalling Crates' concept. This was to be a fata morgana that persisted until finally disproved by the discoveries of Captain Cook in the second half of the eighteenth century.

For a quarter of a century after the publication of his world chart in 1569, Mercator's cartographic activities continued; he was working on a great atlas at the time of his death in 1594. In the following year, the first edition of the complete work was published under the direction of his son, Rumold. The appearance of the Mercator atlas was anticipated by some 15 years by the *Theatrum Orbis Terrarum* of Abraham Ortelius (Oertel) of Antwerp. Ortelius, the friend and rival of Mercator, is

12 Bagrow and Skelton, *History of Cartography* (Cambridge, Mass.: Harvard University Press), pp. 148, 150.

13 R. V. Tooley, *California as an Island,* Map Collectors' Circle, no. 8 (1964).

FIG. 5.5. (opposite) *World map of 1569 by Gerhardus Mercator, on the projection that bears his name.*

credited with the first uniform bound collection of maps designed especially for this form of publication. Previously sheet maps had been assembled in atlas form, as in the various editions of Ptolemy's *Geographia,* and in bound collections of maps by Antonio Lafreri and others in Italy. Lafreri (Antoine du Pérac Lafréry) was a French engraver who in the middle of the sixteenth century settled in Rome, where he assembled and sold collections of maps from various sources. The title page of his collections featured the mythological figure of Atlas carrying the world on his shoulders, but it is undoubtedly because Mercator used the term "atlas" for a book of maps that it is in use today.[14]

Publication of the atlases of Ortelius and Mercator initiated map and atlas production especially in the Low Countries. The first edition of the *Theatrum,* containing seventy maps, appeared in May 1570; it was immediately successful and was followed by two more editions in the same year. Ortelius employed the compilations of many cartographers (usually one per country with acknowledgement), but had the maps engraved on a uniform format.

To illustrate this cartographic form, we have reproduced the map of England and Wales from Ortelius' *Theatrum* (Fig. 5.6). The original map occupies two leaves with an image size of 15 by 18 1/4 inches and is of intermediate, medium, or chorographical scale. In this respect, it stands in contrast to maps of small or geographical scale (such as Fig. 5.5), and those of large scale (sometimes referred to as topographical scale), which will be considered later. In its several elements, Figure 5.6 is representative of much Renaissance cartography.

The title of the map and the name of the author, Humphrey Lhuyd, a Welsh physician, are contained in an ornate cartouche or inset box. This feature, in so-called strap work, with its escutcheon, supporters, etc., is very typical of the cartography of the time. A bar or graphical scale is in another small cartouche, and in the sea are various ships of the period and two sea creatures. The lettering, particularly the italic lettering with sweeps and swash lines in the water bodies, is decorative and suitable to the style of the map. On the land areas are several prominent symbols: profile "mole hills" or "sugar loaves" representing hills and mountains, with no real differentiation between higher and lower

[14] Gerhardus Mercator, *Atlas sive Cosmographicae meditationes de fabrica mundi et fabricati figura* ("Atlas, or cosmographical meditations upon the creation of the universe, and the universe as created"), 3 tomes (Dusseldorf: A. Brusius, 1595).

FIG. 5.6. (opposite) *Plate from Abraham Ortelius'* Theatrum Orbis Terrarum *(1579), showing England and Wales, by Humphrey Lhuyd.*

ANGLIAE REGNI FLORENTISSIMI NOVA DESCRIPTIO. AVCTORE HVMFREDO LHVYD DENBYGIENSE
Cum Privilegio
Scala miliarium Anglicorum
MARE HIBERNICUM
OCEANVS GERMANICVS
OCEANVS BRITANNICVS
SEPTENTRIO.
MERIDIES.
OCCIDENS.
ORIENS.
Scotiæ pars.
Hiberniæ pars.
WALLIA
Sorlinges insulæ

ranges, and no landforms other than uplands and plains represented; scattered buildings in groups, also in profile, representing urban settlements, with little differentiation between larger and smaller places; and rivers widening from source to mouth, often with exaggerated estuaries. No latitude and longitude indications appear, though there are grids on smaller scale maps in the *Theatrum.* The tonal quality of Figure 5.6, resulting from the subsequent hand coloring of the engraving is, in this instance, tastefully executed, but it often detracts from the quality of the work. The stippling in the sea is on the original plate and the coastlines produce a recognizable outline of the country but, understandably, they lack the rigor of a modern controlled survey map. Coastlines are everywhere emphasized by horizontal shading, but areas that are not the prime subject of the map (parts of Ireland, Scotland, and France, in this case) usually have only major cities and landmarks identified. (These areas are detailed in separate maps in the atlas.) Orientation is indicated on the four sides of the map by naming the cardinal directions. The whole is enclosed with a border that gives the effect of a moulding or picture frame; indeed, many such maps have been removed from atlases and are now used as pictures, lampshades, placed under glass on coffee tables, etc. On the reverse side of one leaf of the hinged sheet is a description of the area mapped; such descriptions typically accompany these works.

Between its appearance in 1570 and the last printing, dated 1612, the *Theatrum* was published in more than forty editions and translated into Dutch, German, French, Spanish, Italian, and English. This period in cartographic history has been called the age of atlases because the work of Ortelius and Mercator inspired many others to engage in this lucrative trade—Hondius, Blaeu, Visscher, and Jansson being among the best known.[15] The details of these enterprises need no further explanation; although the productions were sumptuous, generally little cartographic progress was made because these successors of Ortelius and Mercator

[15] Collectors and librarians have prized the sixteenth and seventeenth century productions of the cartographic houses of the Low Countries. These are often treasured as much for their aesthetic appeal as for their geographical content. For a general discussion of this topic, see Lloyd A. Brown, *The Story of Maps* (Boston: Little, Brown, 1949), esp. "The Map and Chart Trade," chap. 6, pp. 150–79. This history of map-making emphasizes earlier phases of the topographic and hydrographic traditions. Arthur L. Humphreys, *Old Decorative Maps and Charts* (London: Halton & Truscott Smith, 1926); this was revised with a new text by R. A. Skelton as *Decorative Printed Maps of the 15th to 18th Centuries* (London: Staples Press, 1952). These last two volumes contain many handsome illustrations including color plates. The writings of Cornelis Koeman on Renaissance cartography in the Netherlands should be consulted.

FIG. 5.7. (opposite) *Bruges from* Civitates Orbis Terrarum *(1572), by Georg Braun and Frans Hogenburg.*

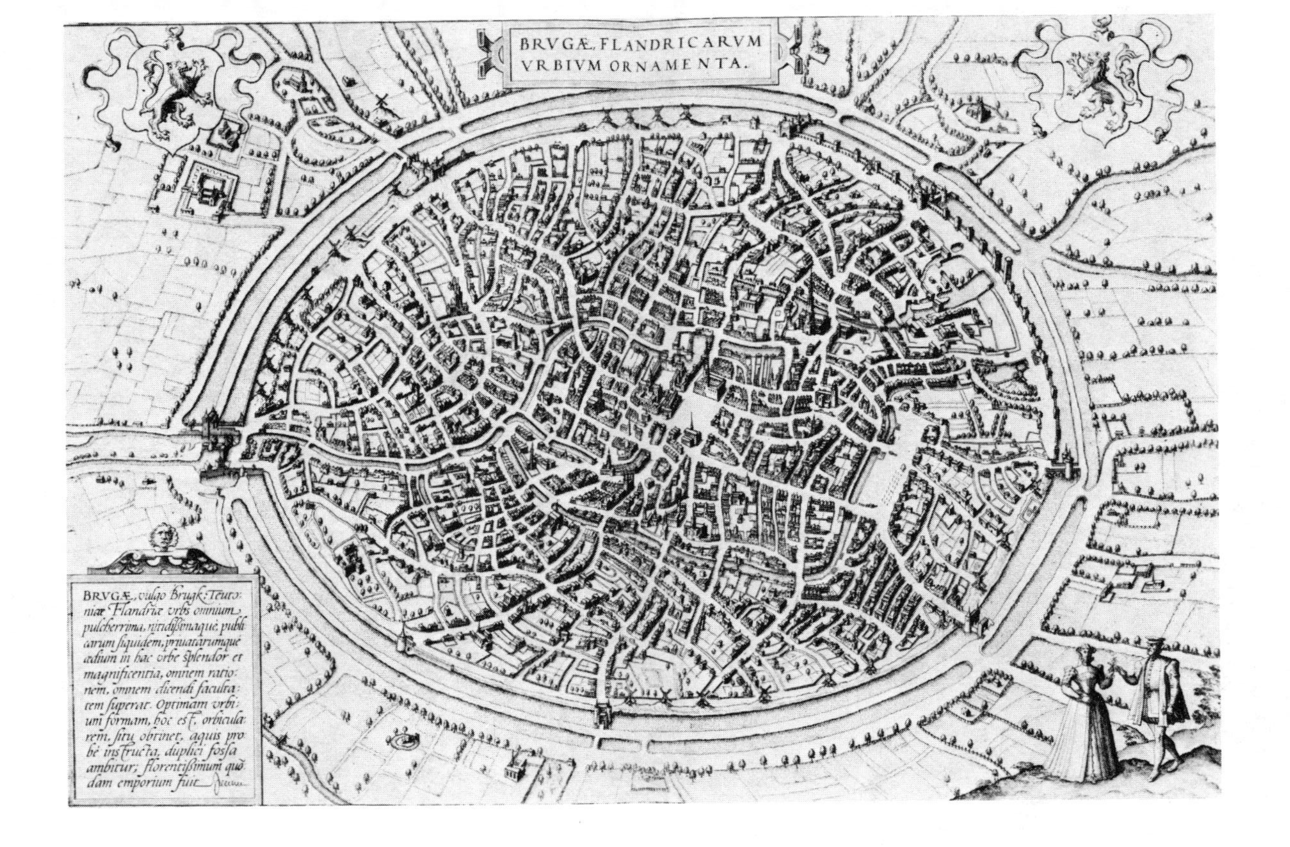
BRVGÆ, FLANDRICARVM
VRBIVM ORNAMENTA.
BRVGÆ, vulgo Brugk: Teuto-
niæ Flandriæ urbs omnium
pulcherrima, nitidißimaque; publi-
carum siquidem, priuatarumque
ædium in hac orbe splendor et
magnificentia, omnem ratio-
nem, omnem dicendi faculta-
tem superat. Optimam urbi-
um formam, hoc est, orbicula-
rem, situ obtinet, aquis pro-
bè instructa, duplici fossa
ambitur; florentißimum quo-
dam emporium fuit.

were often content to reprint the plates with minimal or no amendment. Thus, out-dated plates from the Low Countries were used in Moses Pitt's *English Atlas* published at Oxford as late as 1680–83.[16]

Ortelius included a supplement of historical maps, the *Parergon,* in the 1579 edition of the *Theatrum*—the first edition printed by the famous Plantin (Plantijn) Press. Between 1572 and 1618, an atlas of city plans titled, *Civitates Orbis Terrarum* was published in Cologne by George Braun (Joris Bruin) and Frans Hogenberg. Hogenberg had been employed by Ortelius as an engraver and, in conception, but not in subject, the *Civitates* resembles the *Theatrum.* An example of the contents of the Braun and Hogenberg work is included as Figure 5.7. This oblique or three-quarter view of Bruges shows a city that has undergone little change in its basic morphology since the engraving was made in the sixteenth century. Other cities represented (such as London, which was later devastated by the Great Fire) have altered almost beyond recognition. This cartographic form was also imitated but not materially improved for many years. In our present age, with its emphasis on urban affairs, sources like the *Civitates,* which contains views of Mexico City and Cuzco as well as cities of Europe, Africa, and Asia, are of particular interest.

A discussion of early seventeenth century map-making is incomplete without some reference to the reproduction techniques in use at the time.[17] It has been noted that the first map printed in Europe was a woodcut, as was Waldseemüller's world map (Fig. 5.4). This method of reproduction soon proved too crude for the fine detail on maps, and gave way to engraving with the use of metal, particularly copper plates. Nevertheless, woodcut better lends itself to color reproduction than copper engraving, and an early effort at three-color printing appears in the 1513 edition of Ptolemy, edited by Mathias Ringmann and published in Strassburg. It later became the custom to hand-color the prints of engravings, as in the case of Figure 5.6, a practice that prevailed until the end of the nineteenth century. Understandably, map coloring became an important activity in various cartographic centers, and ladies, sometimes those socially prominent, often engaged in this work.

[16] E. G. R. Taylor, "'The English Atlas' of Moses Pitt, 1680–83," *The Geographical Journal,* vol. 95, no. 4 (1940), 292–99.

[17] *The Penrose Annual, 1964,* Herbert Spencer, ed., devoted several sections to map reproduction; for consideration of historical aspects of this activity, see R. A. Skelton's section, "The Early Map Printer and his Problems," pp. 171–86.

The dissemination of knowledge made possible by the invention of printing in Europe in the last decades of the fifteenth century was an important factor in the spectacular increase in scientific activity that followed. With the publication of *De revolutionibus orbium cœlestium,* in 1543, Nicolaus Copernicus (Niklas Koppernigk, Mikolaj Kopernik) revived the heliocentric theory of the universe, proposed centuries earlier by Aristarchus. Although a greater theorist than observer, Copernicus ushered in a period in which, more than ever before, experimentation was related to observation.[1] A concomitant development was the invention of new instruments and the improvement of existing ones. This led not only to an increase in accuracy, but also extended the range of observation. Progress was made on a number of fronts, some of which affected cartography, directly or indirectly.

As indicated before, the resulting maps are of more concern here than the methods used in their construction, but a few important technological milestones in mapping will be discussed. Triangulation, the fixing of places by intersecting rays, was described by Gemma Frisius in 1533; the plane table, with sighting rule on the drawing surface, which enabled the map to be made at the same time that the angles were drawn, was reported by Leonard Digges in 1571; various tables were compiled, including ephemerides and logarithms —the former by Regiomontanus, Johannes Müller (1436–76), and his student-patron, Bernhard Walther (1430–1504), and the latter developed by John Napier (1550–1617) and Henry Briggs (1561–1630); the pendulum clock, allowing more accurate determination

1 A. R. Hall, *The Scientific Revolution 1500–1800* (London: Longmans, Green and Co., 1954).

Cartography in the Scientific Revolution and the Enlightenment

6

of longitude at fixed points of observation, proposed by Galileo, was built by Christian Huygens in 1657. In addition, the seventeenth and eighteenth century surveyor had available to him the odometer, the magnetic compass, and its derivative, the theodolite. The theodolite was improved toward the end of the period to enable horizontal and vertical angles to be measured simultaneously (altazimuth theodolite).

Although remembered primarily for his work in physics and astronomy, Galileo Galilei (1564–1642) also made maps. Galileo learned of the development of telescopic lenses in the Netherlands and, in 1609, constructed telescopes himself. He was the first scientist to employ the telescope for research purposes and, with a 30x instrument, made what are presumably the first lunar charts by this means. Galileo's original maps were destroyed, along with many other of his works, but several engravings of his lunar drawings are found in his *Sidereus Nuncius* (1610).[2] The lunar sketch map (Fig. 6.1), though crude, was the first to show craters, which Galileo attempted to measure, and seas (maria), which he apparently realized were not water bodies. The fact that a celestial body, like the moon, was a less than ideal form as revealed by Galileo had most important theological implications. Through a "remote sensing" device Galileo first mapped a surface that men would not explore directly for more than 350 years. Galileo discovered two of the moon's librations and initiated the serious study of selenography through his maps and writings.

The work and life of Galileo most dramatically illustrate the spirit of the Scientific Revolution but other, later scientists contributed more to cartography specifically. After Galileo, a generally more favorable attitude toward science existed in Europe, particularly north of the Alps, as evidenced by the foundation of societies to foster scientific inquiry. We can recognize several directions in which cartography moved in the seventeenth and eighteenth centuries, namely the topographic, hydrographic, and thematic mapping traditions.

These three traditions will be emphasized here because it is in these areas that the most important progress was made, but it must not be assumed that there were not interesting developments in other aspects of cartography. For example, the political economist Dr. (later Sir) William Petty as Surveyor General of Ireland (1655–56) under the Protectorate, initiated geometrical surveys of military warrant lands that were recorded on cadastral maps; Petty's influence will be discussed later

[2] Galileo Galilei, *Sidereus Nuncius* (Venetiis: Apud Thoman Baglionum, 1610); and Judith A. Zink (Tyner), "Lunar Cartography: 1610–1962" unpublished M. A. thesis (University of California, Los Angeles, 1963), and by the same author, (now Judith Tyner) "Early Lunar Cartography," *Surveying and Mapping*, vol. 29, no. 4 (1969), 583–96.

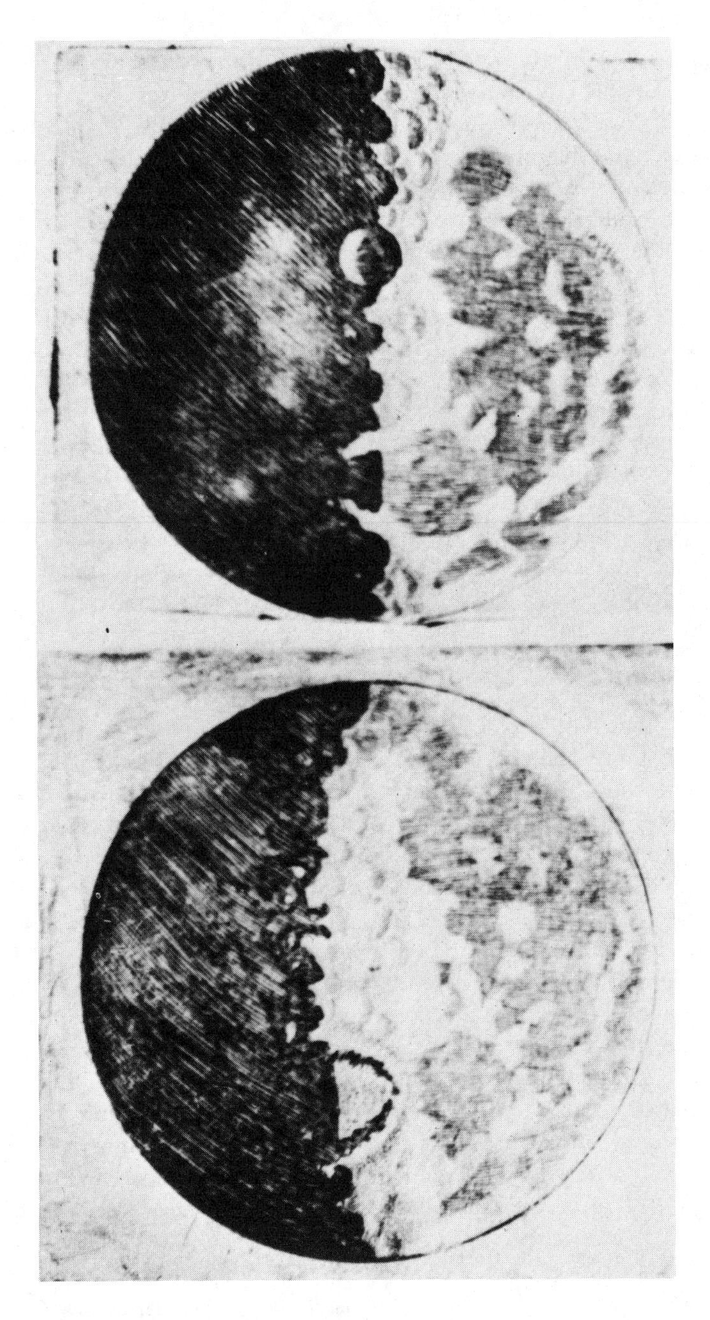

FIG. 6.1. *Engraving of a lunar sketch map by Galileo Galilei from his* Sidereus Nuncius *(1610).*

in this study. In 1675, John Ogilby, the Scottish cartographer who was King's Cosmographer and Geographic Printer under Charles II, published his *Brittania* (Volume I); this work contained strip road maps of the major post highways from London to the main provincial towns with measured distances between places indicated. These are seventeenth century examples of two major cartographic forms—property and route maps—that we have encountered earlier, and to which we will refer again. Globe making (terrestrial and celestial) of this period is exemplified by the work of the Venetian, Vincenzo Coronelli (1650–1718).[3] But let us now turn to thematic maps, perhaps the type of map most neglected by historians of cartography.

A *thematic map* is designed to serve some special purpose or to illustrate a particular subject, in contrast to a general map on which a variety of phenomena (landforms, lines of transportation, settlements, political boundaries, etc.) appear together. The distinction between general and thematic maps is not altogether sharp, but the latter type use coastlines, boundaries, and places (base data) only as points of reference for the phenomenon being mapped (map data or theme of the map) and not for their own sake. One of the most significant contributors to thematic mapping was the English astronomer, Edmond Halley (1656–1742) best known for his prediction of the periodic return of the comet which bears his name. Of course Halley was not the first to make thematic maps.[4] We have mentioned the historical maps in Ortelius' *Theatrum;* we know that Oronce Finé, a well-known French cartographer, made Biblical maps (now lost) in the mid-sixteenth century; and earlier examples could be cited. But these are very different from the thematic maps of Halley, who illustrated a number of his own scientific theories by cartographical means.

Edmond Halley, whose mapping endeavors well exemplify the cartography of the Scientific Revolution, was a Fellow and, for a time, clerk to the two Secretaries of the then youthful Royal Society of London. Through this connection he became acquainted with many of the greatest scientists of his age including: Johannes Hevelius of Danzig, the foremost lunar cartographer of the seventeenth century; Giovanni Domenico (later Jean Dominique) Cassini, who supervised the construction of an earth map (*planisphère terrestre*) on the floor of the

[3] The memory of Coronelli is honored in the name of an international society: Coronelli—World League of Friends of the Globe.

[4] Norman J. W. Thrower, "Edmond Halley and Thematic Geo-Cartography," *The Terraqueous Globe* (Los Angeles: William Andrews Clark Memorial Library, University of California, Los Angeles, 1969), pp. 3–43, and "Edmond Halley as a Thematic Geo-Cartographer," *Annals of the Association of American Geographers,* vol. 59, no. 4 (1969), 652–76.

Paris Observatory; and Sir Isaac Newton who contributed specifically to cartography and geodesy through his announcement, before it was proved, that the earth is an oblate (polar flattened) spheroid rather than a prolate (equatorial flattened) one.

Halley's first significant cartographic venture was a celestial planisphere (star chart) of the constellations of the Southern Hemisphere made during a stay of about a year at St. Helena and published in 1678. Some years after this Halley became editor of the *Philosophical Transactions,* in which he published his first important terrestrial map in 1686 (Fig. 6.2). This map, which accompanies an article by Halley on the trade winds, is considered the first meteorological chart.[5] It well illustrates a new trend in cartography, focusing, as it does, on a single physical theme—the direction of the prevailing winds in the lower latitudes. Halley selected a suitable projection (the Mercator) for showing the phenomenon being mapped (the map data). The grid is composed of 10 degree lines of latitude, and 15 degree lines of longitude (based on the Prime Meridian of London) which each represent 1 hour of earth rotation. No extraneous decoration appears on this chart. The prevailing winds are shown generally by tapering strokes, the tails indicating the direction from which the wind usually comes. Halley used arrows, which are now the conventional symbols for such phenomena, only in the Cape Verde area. A minimum of base data needed for showing the thematic distribution appears on the map (some may even object to it being so designated since it lacks a title and scale indication). The addition of the north and west coast of Australia to the world map resulted largely from the sixteenth and seventeenth century explorations of the Dutch in this area. The map below Halley's wind chart shows coastlines from a present-day source drawn on the same projection for comparison.

In 1698 Halley was given a temporary commission as a Captain in the Royal Navy for the purpose of investigating the earth's magnetic field. He took command of a small ship, the *Paramour,* and embarked on a voyage sponsored by the Royal Society, which has been called "the first sea journey undertaken for a purely scientific object."[6] Shortly after his return to England in 1700, Halley published a map which he titled "A New and Correct Chart shewing the Variation of the Compass in the Western and Southern Ocean" (Fig. 6.3). The map uses isogones (lines of equal magnetic declination, i.e., variation in degrees from geographical north) and is the first such printed map extant, and also the

5 Sydney Chapman, "Edmond Halley as Physical Geographer and the Story of His Charts," *Occasional Notes of the Royal Astronomical Society,* no. 9 (London: 1941).

6 *Ibid.,* p. 5.

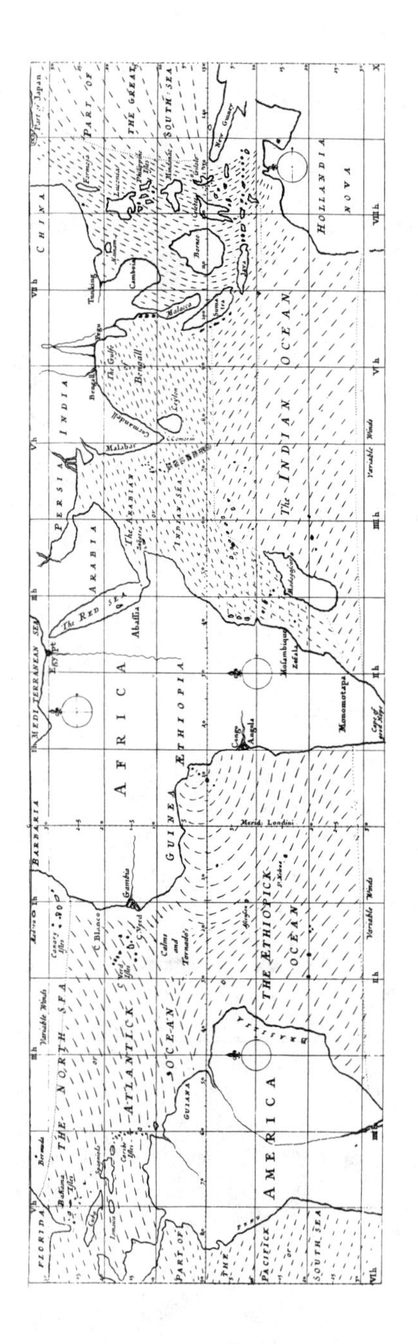

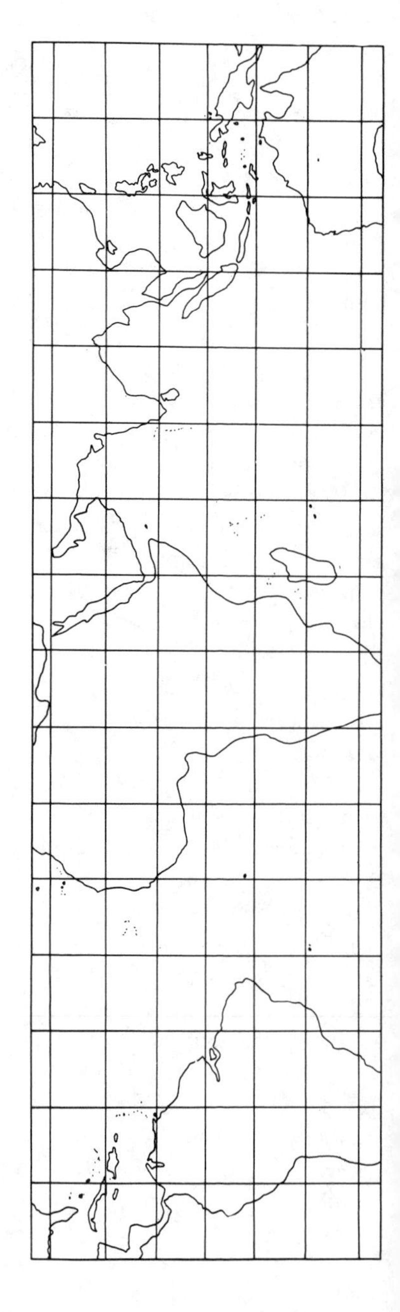

FIG. 6.2. (opposite) *Chart of the trade winds by Edmond Halley (1686) from the* Philosophical Transactions *(above); (below), a map on the same scale and projection to indicate a modern delineation of the world's coastlines, for comparison.*

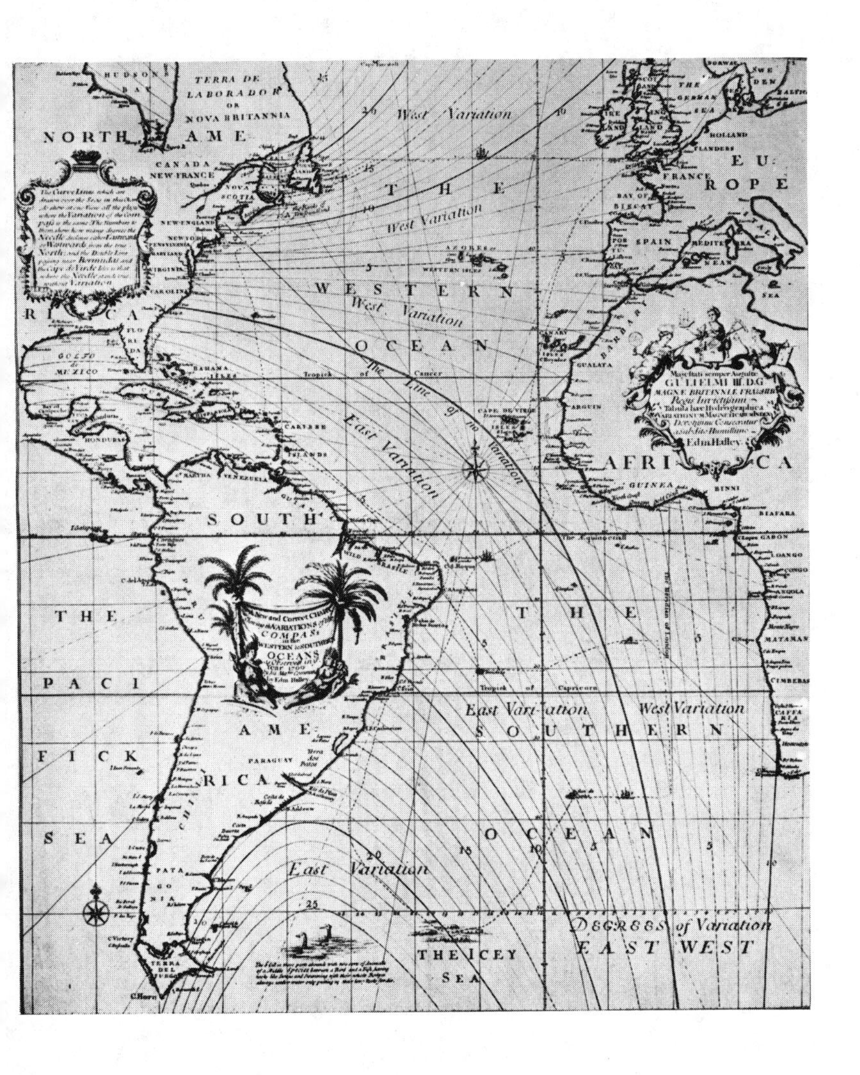

FIG. 6.3. (above) *Isogonic map of the Atlantic (1701) by Edmond Halley.*

first published isoline map of any kind.[7] In thematic, quantitative cartography isolines (isarithms) are lines connecting points having equal intensity of phenomena, which have transitional degrees of intensity. Thus coastlines, which normally and theoretically depict the same intensity, cannot be considered isolines in this sense. (An exception, of course, would be when high and low tide lines are shown on the same map.) We shall encounter this fundamental means of cartographic representation in other connections later in this work (Appendix B). In 1702, Halley published a larger map which extended the isogones to the Indian Ocean (based on the observations of others) but not to the Pacific, for which no adequate data were yet available. Both Halley's Atlantic and World isogonic charts feature rococo-style cartouches, but generally have less decoration than is usual for this period.

A third map by Halley (Fig. 6.4) links the thematic with the *hydrographic tradition*; it again illustrates the interest of this scientist (whom his contemporaries considered second only to Newton among the English natural philosophers) in cartographic work. Before discussing this map in particular, we should review, briefly, the progress of marine cartography up to this point. The delineation of coastlines (separating those two fundamental geographical quantities, land and water) had, of course, been part of cartography since antiquity; and in the portolan chart of the later Middle Ages, we have a cartographic genre specifically designed for the navigator. We know that this type of chart formed the early basis of Iberian mapping but, in time, as the European view of the world expanded through geographical discoveries, the character of the sea chart also underwent change. To assist seamen, various navigational manuals were published, some of which contained views of coastal features and, later, charts. We have discussed the contributions in the sixteenth century of northern Europeans (especially the German and Low Country cartographers) to mapping, the high point of which was the development of a chart of particular use to the navigator—the Mercator. A work comparable in scope in its field to Ortelius' *Theatrum* or Braun and

[7] Werner Horn, "Die Geschichte der Isarithmenkarten," *Petermanns Geographische Mitteilungen,* vol. 53 (1959), 225–32. However, a simple isogonic map in manuscript form by a Jesuit Padre of Milan, Christoforo Borri, is reported in the writings of Athanasius Kircher. Kircher apparently worked on an isogonic map himself which he intended to include with his *Magnes sive de arte magnetica opus tripartitum* (Rome: 1643); Halley refers to Kircher in his writings.

FIG. 6.4 (opposite) *Chart of the tides in the Channel (1702) by Edmond Halley.*

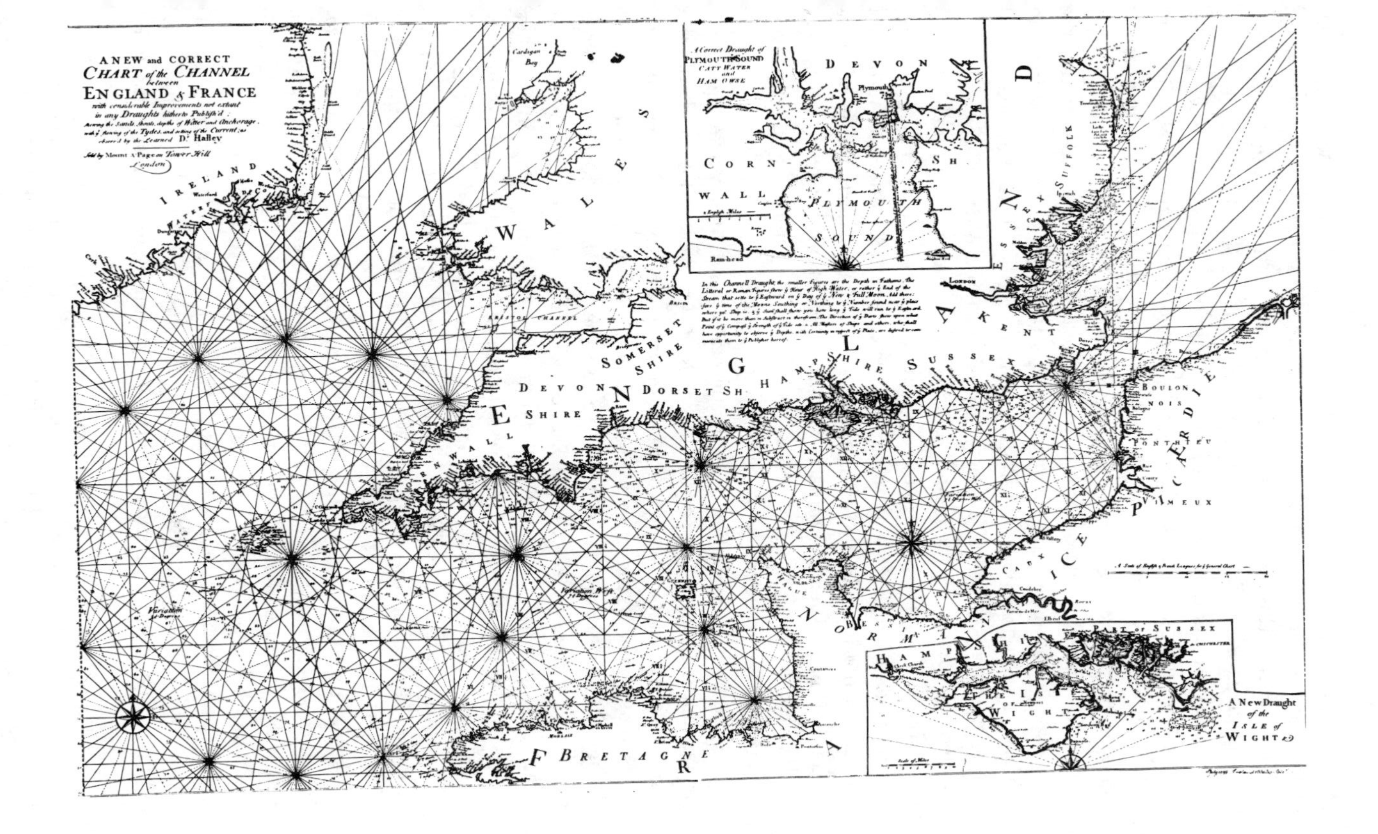

A NEW and CORRECT
CHART of the CHANNEL
between
ENGLAND & FRANCE
with considerable Improvements not extant
in any Draughts hitherto Publish'd;
shewing the Sands, Shoals, depths of Water and Anchorage,
with ye flowing of the Tydes, and setting of the Current; as
observ'd by the Learned Dr. Halley
Sold by Mount & Page on Tower Hill
London
A Correct Draught of
PLYMOUTH SOUND
CATT WATER
and
HAM OWSE
DEVON
CORNWALL
PLYMOUTH SOUND
A New Draught
of the
ISLE of
WIGHT
PART OF SUSSEX
HAMPSHIRE
WALES
IRELAND
Cardigan Bay
BRISTOL CHANNEL
SOMERSET SHIRE
DEVON SHIRE
DORSET SH
HAMPSHIRE
SUSSEX
KENT
SUFFOLK
LONDON
ENGLAND
BRETAGNE
NORMANDIE
CAUX
PICARDIE
BOULONNOIS
PONTHIEU
VIMEUX

Hogenberg's *Civitates,* was the first true printed atlas of sea charts, *De Spieghel der Zeevaert* (1584–85) by Lucas Janszoon Waghenaer. So influential was this work that an English translation of the "waggoner" appeared as *The Mariners Mirrour* as early as 1588, the year of the attempted invasion of England by the Spanish Armada. However, no self-respecting maritime country could for long be dependent upon another nation for charts—partly because a great deal of secrecy surrounded such cartographic work and the chart-makers supplied foreigners only with the information they chose to disclose. Thus England entered the field of chart-making, the theoretical basis being provided by Tudor and Stuart mathematicians including Edward Wright (mentioned earlier in connection with Mercator's chart), while the actual surveying was accomplished by practical men such as William Borough.[8] French marine surveyors, who also were active at this period, were to make striking improvements in chart-making in the seventeenth century, as indicated later. It is the French who eliminated from marine charts the wind roses which had been a feature of such maps since the portolan was introduced, and which were no longer necessary as new instruments and improved navigational techniques were developed.

Halley, who combined both practical and theoretical qualities, began marine surveying a decade before he took command of the *Paramour.* He made a chart of the mouth of the River Thames and five years later, one of the Sussex coast. Upon returning from the Atlantic voyage, Halley secured permission to use the *Paramour* for the purpose of surveying the English Channel, and in 1702 published a map resulting from this activity (Fig. 6.4). Superficially, this map resembles other charts of the period with its representation of coastlines (understandably much improved over those of the same area in Ortelius' atlas of a century and a quarter earlier), shoals, anchorages, depths (marked in fathoms), etc. Even the compass roses with radiating lines are reminiscent of earlier marine charts, but it differs from these in at least two particulars. First, Halley provides a formula for estimating the height of the tides at certain places, which are indicated on the map by Roman numerals (the direction of tides being shown by arrows). Second, Halley took angles by the sun for the greater accuracy this method provides rather than by magnetic compass as was the usual practice at that time. Halley gave up his active naval command at about the time of the publication of his tidal chart but continued to interest himself in

[8] Eva G. R. Taylor, *The Mathematical Practitioners of Tudor and Stuart England* (Cambridge: Cambridge University Press, 1954); and William Bourne, *A Regiment for the Sea* (London: The Hakluyt Society, Second Series, no. 121, 1961). See also Adrian H. W. Robinson, *Marine Cartography in Britain* (Leicester: Leicester University Press, 1962).

cartography to the end of his life. In 1715 he produced a map of the shadow of the moon over England, resulting from the total eclipse of the sun; the time taken by the passage of the moon's shadow is also indicated. This map beautifully illustrates the concern of Halley, who became Astronomer Royal at Greenwich Observatory in 1720, with the entire cosmos. Since it was made before the event it depicts, this map demonstrates the highest attribute of science—the ability to predict. Like his other major thematic maps, it is a work of great originality.

We have discussed the isogonic maps of Halley and indicated that his Atlantic chart of 1701 is, presumably, the earliest published isoline map. However, two manuscript maps with isobaths (lines of equal depth of water) are known to antedate Halley's map. They are by Pieter Bruinss, (1584) and Pierre Ancelin (1697); apparently no cartographer between these dates thought of joining up points of equal depth. A scattering of depth values along coasts, a feature of the maps of the Renaissance cartographers, was made possible by that age-old navigational instrument, the lead and line for plumbing depths. However, in 1729, an engraved map of isobaths was published (Fig. 6.5). It was the work of Nicholas Samuel Cruquius (Cruquires), a Dutch engineer, and shows depths in the Merwede River, a distributary of the Rhine.[9] Depth values are marked in the main river but in the smaller tributaries and other water bodies, the usual unquantified form lines of the period are drawn (compare Fig. 6.5 with Fig. 6.8). Shortly after this time, in 1737, a generalized isobathic chart of the Channel was made by the French cartographer, Philippe Buache.[10] In the second half of the eighteenth century there was great interest in charting the coasts of the world, particularly for those areas only recently discovered by Europeans. Outstanding charts of New Zealand and part of Australia, and North America were produced by Captain James Cook. His work on the Pacific coast of North America was continued by George Vancouver who had accompanied Cook on his second and third voyages of discovery. Official government hydrographic departments were founded in France in 1720 and in Britain in 1795, though in the latter case contemplated decades earlier.[11]

It is understandable that isobaths, which show the configuration of a surface (in the case of Cruquius' map a river bed—a rather different

[9] S. J. Fockema Andreae and B. van't Hoff, *Geschiedenis Der Kartographie Van Nederland* ('s-Gravenhage: Martinus Nijhoff, 1947).

[10] R. A. Skelton, "Cartography," in Charles J. Singer, ed., *A History of Technology* (Oxford: Clarendon Press, 1954–58), vol. 4, chap. 20, pp. 596–628. Buache's generalized map is reproduced on p. 613 of this volume.

[11] For a popular account of Britain's important role in hydrographic mapping see Sir John Edgell, *Sea Surveys* (London: Her Majesty's Stationary Office, 1965); the author served as Hydrographer of the Navy from 1932 to 1943.

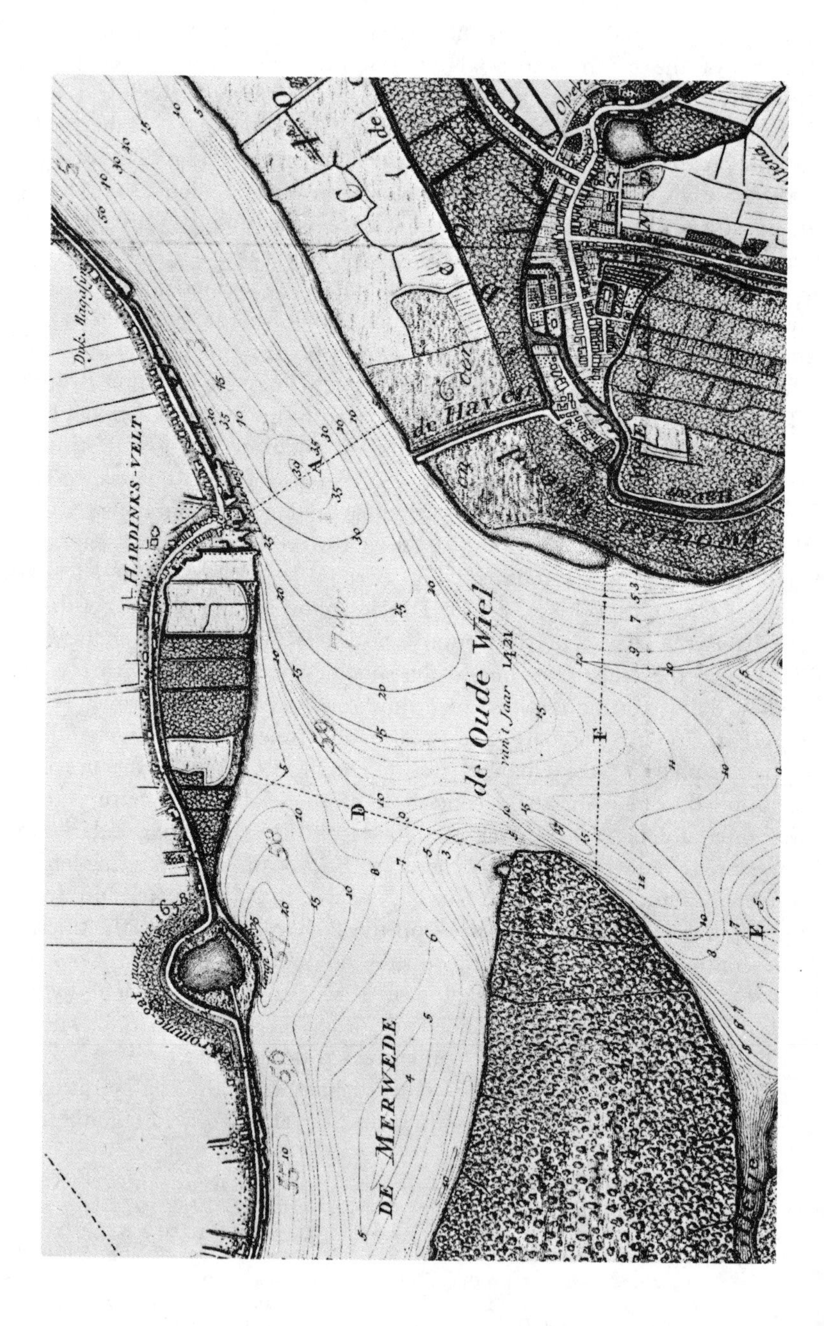
HARDINKS-VELT
de Oude Wiel
van 't Jaar 1421
DE MERWEDE
de Haven

concept than isogones) should have been the first form of the contour lines applied to delineate any part of the lithosphere. The surface of the water, though variable in height through time within certain limits, forms a convenient and natural datum (base to which measurement can be referred) and the lead and line provides an easy means of gathering the needed depth information. It was several decades before the contour principle was applied in a significant way to dry land surfaces, partly because of the difficulty of making the necessary measurements, and partly because of the preference on the part of cartographers and map users for the hachure technique of landform representation which will be discussed subsequently.

During the seventeenth and eighteenth centuries, France became the leader in *topographic mapping,* developing methods which became standard and, later on, were widely adopted elsewhere. This began after the astronomer Giovanni Domenico Cassini (1625–1712), who was a professor at Bologna, accepted an invitation to the *Académie Royale* in Paris. This scientific society served a similar function in France to that of the Royal Society in England; both societies were founded in the mid-seventeenth century and both concerned themselves with a wide variety of scientific problems, including mapping and charting.

Long before this time the mapping of countries and smaller land areas was, of course, undertaken. We have illustrated this with Humphrey Lhuyd's map of England and Wales from Ortelius' *Theatrum* (Fig. 5.6). More detailed maps of this area—the county maps of Christopher Saxton (1542–1606) and John Norden (1548–1626) and estate maps or "terriers" by a number of cartographers—had also appeared. Examples could be provided from a number of European countries where progressively larger scale and more detailed maps were made so that by the end of the sixteenth century regional maps, some consisting of multiple sheets, existed for a good many areas.[12] In some instances, e.g. the work of Willebrord Snell (Snellius), 1591–1626, who developed the method of determining distances by trigonometric triangulations, real progress was made in accuracy through better techniques and improved instruments. However, if maps of even larger and uniform scale covering an extensive area were to be made suitable for administrative, engineering, and military purposes (topographic quadrangles), more rigorous standards had

[12] Bagrow and Skelton, *History of Cartography* (Cambridge, Mass.: Harvard University Press), esp. pp. 143–76.

FIG. 6.5. (opposite) *Section of an isobathic chart of the Merwede River (1729), by Nicholas Samuel Cruquius.*

to be applied. Cassini's arrival in France in 1669 initiated the topographic survey of that country.[13]

We have already alluded to the *planisphère terrestre* laid out on the floor of the Paris Observatory by Cassini; interestingly, Halley visited Cassini at the Observatory in 1682 when he was engaged in this work. Cassini's master map of the world (with an azimuthal projection centered on the North Pole) recalls the *Padrón General* of the *Casa* in Spain or Ptolemy's maps in that an attempt was made to collate current geographical data. But, unlike these works in which the location of places depended to a large extent on verbal information from travelers by land or upon dead reckoning estimates supplied by sailors, no place was added to Cassini's planisphere unless its position had been determined astronomically. This information was published in map form by Cassini in 1696 but it also became well-known through the compilations of Guillaume De Lisle (1675–1726) and other cartographers who had access to the master map in the Paris Observatory. In this way the true length of the Mediterranean, approximately 42 degrees (which we have seen was accurately determined by Arab astronomers by the twelfth century), was, as far as we know, first correctly recorded on printed maps. The locations of other areas were similarly improved and conjectural information, especially on the interiors of continents, eliminated. It was this imaginary cartography that had evoked Jonathan Swift's well known satirical comment:

> So Geographers, in Afric-maps,
> With savage-pictures fill their gaps;
> And o'er unhabitable downs
> Place elephants for want of towns.
> [*On Poetry,* line i. 117]

A detailed and accurate map of France in multiple sheets and employing uniform standards and symbols was needed; at the request of Colbert, and with royal support, the *Académie* under Cassini attempted to meet this challenge. The first step was to measure the arc of the meridian of Paris to ascertain the length of a degree of latitude. This was undertaken by the Abbé Jean Picard by means of triangulation, a method of which he was a strong advocate. Picard used a quadrant with telescopic sight and filar micrometer of his own invention. The work completed in 1670 was the basis for a series of nine topographic sheets of the Paris area made by Du Vivier. Meanwhile, Picard and a group of engineers,

[13] Sir George H. Fordham, *Some Notable Surveyors and Map-Makers of the Sixteenth, Seventeenth, and Eighteenth Centuries and Their Work* (Cambridge: Cambridge University Press, 1929), esp. chap. 3. See also Lloyd A. Brown, *Jean Dominique Cassini and His World Map of 1696* (Ann Arbor: University of Michigan Press, 1941).

using fixed points provided by the astronomers of the *Académie,* surveyed the coasts of France. Great discrepancies were found between the coastlines established by this marine survey and their representation on the best previous maps of France, those of the Sanson family, as shown in Figure 6.6. It is to a member of this family, Nicholas, that we owe the development of the important Sinusoidal equal area projection, also known as the Sanson-Flamsteed projection. The second name is attached because John Flamsteed, who preceded Halley as Astronomer Royal, used the projection for a star chart.

Detailed charts arising from the new French marine survey drawn on the Mercator projection and with latitude and longitude indications were published later as *Le Neptune François, ou Atlas Nouveau des Cartes Marine* (1693). Picard died before this work was completed, but subsequently the meridian of Paris was extended from the Channel to the Pyrenees. A practical result of this was that, with extension, the triangulation could be the basis of more accurate topographic maps of an entire country than had ever been produced before. In the more theoretical realm, measurement of the degrees of latitude over a long north-south line cast doubt, ironically, upon Picard's belief in a perfectly spherical earth and on the prolate spheroid theory supported by Cassini's son, Jacques, and others. However, further measurements sponsored by the *Académie* in equatorial South America and in Lapland confirmed the general correctness of Newton's hypothesis of the earth being an oblate spheroid.

After the death of his father, the survey of France was continued by Jacques Cassini de Thury, who also succeeded his father as the head of the Paris Observatory. The meridian was resurveyed with refined methods and triangulation extended east and west. Figure 6.7 shows a small section of the map of the triangulation of France by Giovanni Maraldi and Jacques Cassini (1744), the whole work consisting of some 40,000 triangles. Cassini was assisted by his son, César François, who carried the topographical mapping of France, including the filling in of detail, to a virtual conclusion by subscription, and without official financial support. The few sheets not completed before his death in 1784 were prepared under the direction of his son Jean. The topographical map of France—182 sheets on the scale of "*une ligne pour cent toises,*" (1:86,400) was eventually completed in 1793. Thus four generations of the Cassini family over a period of more than 100 years had been involved in the first true topographic survey of an entire country in which the principle of providing a rigorous framework for the whole survey before the details were filled in, was applied. The sheets, when all assembled, measure approximately 36 × 36 feet.

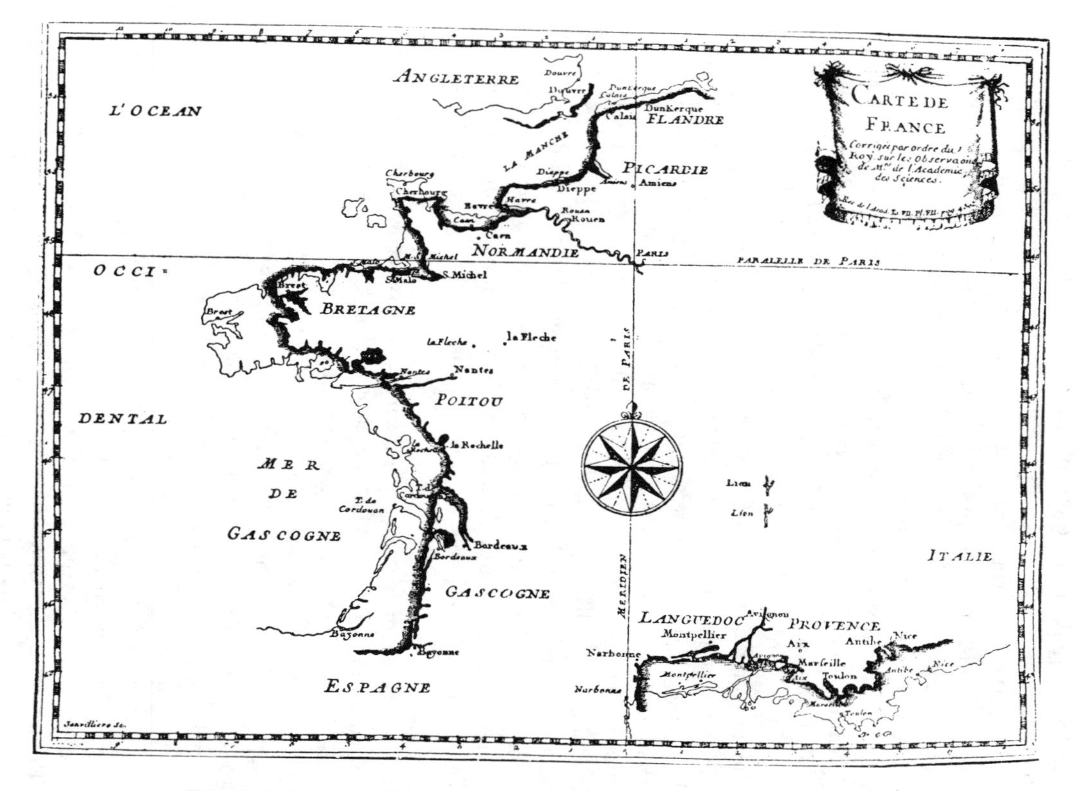

FIG. 6.6. *The rendering of the coastlines of France resulting from surveys of the scientists of the* Académie Royale, *1693 (shaded) superimposed over the delineation of Sanson, 1679 (line).*

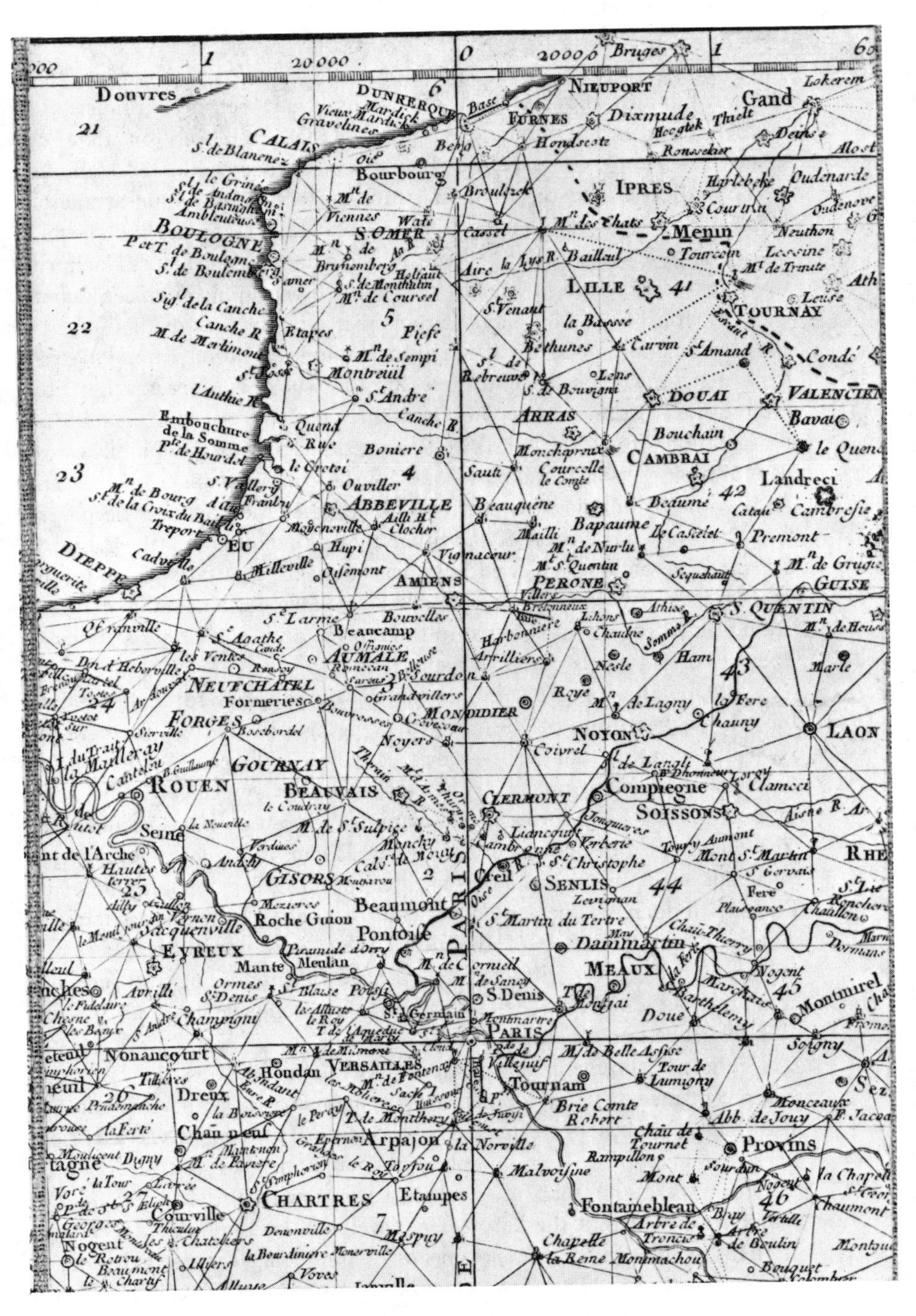

FIG. 6.7. *Section of the map of the Triangulation of France by Giovanni Maraldi and Jacques Cassini (1744).*

Figure 6.8 is a small sample of one of these well-engraved topographical maps, showing part of the northeast coast of France. The symbolization is essentially the same as that employed on the Paris sheets of over a century before. In particular, the terrain representation which indicates two or, at most, three surface levels is unsatisfactory. Hachures (i.e., short lines the thickness of which indicates the steepness of the slope) were used to delineate these tabular surfaces. Hachuring was later systematized in 1799 by the Saxon topographic engineer, Johann G. Lehmann (1722–1805) but it remained a qualitative method of landform representation in the sense that absolute elevation cannot be read from hachures alone.[14] Hachures are drawn downslope, rather than around the feature as are contours. A three-dimensional effect, simulating illumination from the northwest was produced by emphasizing the hachures on the south and east slopes on the Dufour maps of Switzerland. At best the technique is expressive, but at worst the hachures degenerate into "hairy caterpillars." In addition to the hachures on the *Carte de Cassini* some obliquely viewed landforms appear (e.g., coastal sand dunes on Fig. 6.8). These, and the symbols for smaller settlements and forests, are not in plan view in contrast to the representation of larger settlements and, of course, roads, rivers, coastlines, etc.

Delineation of the continuous three-dimensional form of the land has always been one of the most challenging problems in cartography. As we have seen, from earliest times, so-called "fish scales," "mole hills," or "sugar loaves" in profile or, at best, oblique views have been used to represent relief. Such forms utilize planimetric position in two dimensions. The planimetric displacement which results can, apparently, be tolerated on small scale maps but not on those of large or topographic scale. Features hidden by the landforms can, of course, be realigned only by destroying the accuracy of the map. What was needed was a planimetrically-correct quantitative method of terrain rendering and such had actually been developed before this time in isometric lines of the sort used by Halley or Cruquius.

The measurement of elevation of dry land features had been possible ever since the development of trigonometry and the invention of instruments, such as the altazimuth theodolite. However it was laborious work to survey the large number of points needed for an accurate isometric (contour) map of the land surface with this instrument. For a time it seemed that the barometer, developed by Evangelista Torricelli in 1643, might provide an easy means of producing the needed data, as the lead and line had for isobathic maps. Blaise Pascal (1623–62)

[14] Skelton in Singer, *A History,* p. 611 with an illustration of Lehmann's method.

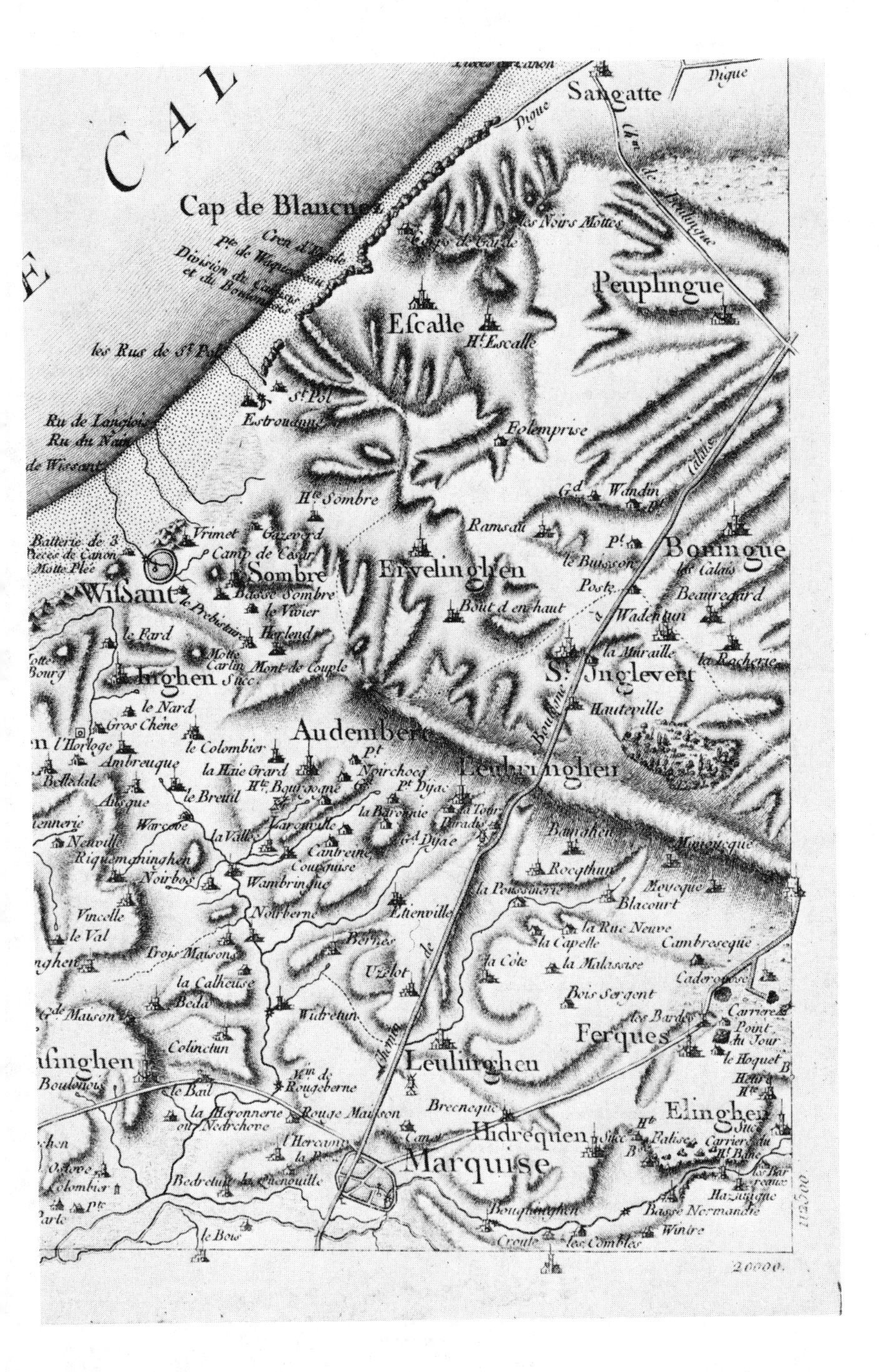

FIG. 6.8. *Section of engraved topographic map of France from the Cassini Surveys.*

demonstrated that pressure decreases as a function of increasing elevation and this principle was employed in cartography by an English physician, Christopher Packe. In the compilation of his Philosophico-Chorographical chart of Kent, published in 1743, Packe used barometric readings converted to elevation figures and plotted these as spot heights from a datum provided by high tide in the Channel.[15] However, in spite of his precaution of taking readings only on days with similar weather, the temporal variation of pressure rendered this method less than satisfactory. On the *Carte de Cassini* some spot heights appear. A number of modest scale attempts to draw contours, i.e., lines connecting all points with the same altitude, were made by military engineers in the second half of the eighteenth century. An interesting early attempt at contouring appears in a map of the terrain of Oxford where from the highest point in the city (Carfax), the contours are referred downward, in the manner of isobaths, rather than upward from a datum such as sea level. The earliest use of contours covering a large area on topographic maps is attributed to a French engineer, J. L. Dupain-[Dupin] Triel (1765–1811). The date of this work is 1791, although he later added tints between selected contour levels to produce a colored hypsometric (altitude tint) map.[16] But it was not until the middle of the nineteenth century that contouring displaced the hachure method of relief representation in many of the great national surveys. In some cases contours were drawn as a basis for hachuring and then removed from the final sheets after the hachures were drawn.

The value of detailed topographic mapping of countries was quickly appreciated by administrators. Very soon after its publication, the *Carte de Cassini* was taken over by the French government. Later, in the hands of Napoleon Bonaparte who had a great appreciation of geography, mapping became a prime instrument of administration and conquest. Meanwhile, in Britain, General William Roy who had long advocated the detailed mapping of that country embarked on a trigonometric survey in 1783. It is from this beginning that the official, quasi-military Ordnance Survey was formed.[17] The British and French triangulation networks were connected across the Channel in 1787. A quarter of a century before this the French survey and the triangulation of the Netherlands by Willebrord Snell, noted earlier, had been joined overland.

[15] Eila M. J. Campbell, "An English Philosophico-Chorographical Chart," *Imago Mundi,* vol. 6 (1950), 79–84.

[16] M. Foncin, "Dupin-Triel [*sic*] and the First Use of Contours," *The Geographical Journal,* vol. 127 (1961), 553–54. For an excellent summary of the French and Low Country contributions, particularly to isobathic and contour mapping, see Françoise de Dainville, "From the Depths to the Heights," *Surveying and Mapping,* vol. 30, no. 3 (1970), 389–403.

[17] R. A. Skelton, "The Origins of the Ordnance Survey of Great Britain," pt. 3 of "Landmarks in British Cartography," *The Geographical Journal,* vol. 78 (1962), 406–30.

Some of the most interesting cartographic developments were soon to take place in areas where there were fewer restraints—away from the main population centers of northwestern Europe. In Ireland, for example, instead of using one meridian, as had been done in France, the Ordnance Survey employed several meridians to further reduce error in mapping. Leveling, also introduced in the surveys of this area, permitted contours to be drawn with greater accuracy and facility. In India, surveying on the European model was begun in Bengal, in 1763, by James Rennell. Triangulation was initiated in Madras in 1802, which led to the Great Trigonometrical Survey of India. Using this framework, the Survey of India eventually produced topographic maps of essentially all of the subcontinent (so that later it could be claimed that India was the best surveyed large country in the world). Much was also learned about earth magnetism as the result of these surveying activities.

One of the results of the French Revolution was the reform of weights and measures. In 1791 the Paris *Académie des Sciences,* formerly the *Académie Royale des Sciences,* defined the meter as 1/10,000,000 of a quadrant of the terrestrial meridian and various countries converted their standard linear measure to the metric system. This led to the concept of the "natural scale" in cartography, whereby one unit of length on the map is represented by a given number of like units on the earth. This so-called representative fraction (R.F.) was first used in France in 1806. The representative fraction, e.g., 1:63,360, is a more general expression of scale than its verbal equivalent, one inch equals one mile. These two scale expressions plus the graphical scale are in common use today.

In the American Colonies there had been a great concern with mapping of the land since the earliest European settlements. In 1585 John White drew a map of the Cape Hatteras area during the second Roanoke expedition of Sir Walter Raleigh. Some years later (1608) Captain John Smith explored the Chesapeake Bay and produced an important map of that locality, on which reliability is indicated in terms of those areas he had visited and those he had "by relation." One of the most important American maps of Colonial times was that produced in 1670 by Augustine Hermann (a Bohemian) for Lord Calvert, of his Maryland patents and a large region beyond. A map covering an even larger area from Virginia to the Great Lakes was made by Joshua Fry (under whom George Washington served in the French and Indian Wars) and Peter Jefferson (father of Thomas Jefferson) in 1751.[18] Both

18 Coolie Verner, with introduction by Dumas Malone, *The Fry & Jefferson Map of Virginia and Maryland* (Charlottesville: University Press of Virginia, 1966).

George Washington and Thomas Jefferson, like many of their contemporaries in America, engaged in surveying and had a great interest in mapping and geographical exploration. The most ambitious map made in Colonial America was that by John Mitchell dated 1755. Like all of those referred to in this paragraph it was printed and published outside of the country and, while adding to geographical knowledge, was essentially European in methodology and symbolism. After the Revolutionary War, American cartography developed along more independent lines. An immediate concern was the settlement of the vast Public Domain and a committee under the chairmanship of Thomas Jefferson proposed a method for its subdivision. (Cadastral mapping of this area and other American cartographic developments will be considered in the next chapter.)

A number of eighteenth century mathematicians turned their attention to the solution of cartographic problems, notably Johann H. Lambert (1728–77).[19] Lambert probably did more than any other individual to advance the study of map projections. He invented several projections which are in use today, including the Cylindrical Equal Area, the Conic Conformal with two standard parallels, the Conic Equal Area with one standard parallel, the Azimuthal Equal Area, and the Transverse Mercator. This last projection, often known as the Gauss Conformal (and thus wrongly attributed to the famous German mathematician Karl F. Gauss, 1777–1855) has recently found important uses and will be discussed subsequently.[20] Rigobert Bonne (1727–95) devised the projection bearing his name in 1779 which shortly became popular. Bonne's Projection is based on a central meridian true to scale with all parallels also being true to scale and concentric circles. This projection, of which many cases are possible, replaced the less satisfactory Cassini Projection for the French topographic surveys in 1803, and thereafter was widely used elsewhere for such purposes.

[19] Information on map projections, including methods of construction, is contained in the standard textbooks on cartography listed earlier, and in a number of special publications. Among the latter, outstanding works are Charles H. Deetz and Oscar S. Adams, *Elements of Map Projections* (Washington, D. C.: U.S. Coast and Geodetic Survey, 1934); Irving Fisher and Osborn M. Miller *World Maps and Globes* (New York: Essential Books, 1944); and James A. Steers, *An Introduction to the Study of Map Projections*, 13th ed., (London: University of London Press, 1962). A famous unsolved mathematical problem concerns coloring of maps. If we assume that: (*a*) two areas that have a strip of boundary in common must have different colors; (*b*) two countries that have isolated points of boundary in common may have the same color; (*c*) each area is one connected piece; then how many colors would be needed to color the map? It has been proved that all maps can be colored with five colors, but no one has produced a map which required five colors.

[20] Johann H. Lambert, *Beyträge zum Gebrauche der Mathematick und deren Anwendung*, 5 parts, (Berlin: Reimer, 1765–72).

These developments were followed by the use of the Gnomonic projection for terrestrial maps in the first years of the nineteenth century by C. T. Reichard of Weimar. On the Gnomonic projection a straight line is a great circle, the shortest distance between two points on the globe. This attribute makes the projection valuable for plotting routes which may be approximated on the Mercator chart with a series of straight line segments (rhumb lines), which can be navigated by compass. In 1805 the Homolographic (equal area) projection was developed by Karl B. Mollweide, whose name is often given to this projection (which is also discussed later in this work). Another equal area projection devised at this time was the Albers Conic with two standard parallels. In spite of the invention of these useful equal area projections for part or all of the world, projections with no prime properties, those which were familiar, or easy to construct, continued to be used for many years for distributional maps.

Some immediate effects of the Scientific Revolution upon cartography were discussed in the previous chapter. The cartographical development of other important concepts propounded in the seventeenth and eighteenth centuries took place during the Industrial (and Technological) Revolution which followed. Thus the foundations of modern statistical methods were laid in the seventeenth century by such workers as Sir William Petty who investigated demographic problems and popularized the study of vital statistics.[1] His contemporaries, Huygens and Halley concerned themselves with the theory of probability which led, in the eighteenth century, to the use of accurate statistical methods. The holding of regular censuses which began, in modern terms, in Sweden in 1749, in the United States in 1790, and in Britain in 1801, provided a large potential source of mappable data. There was also a great increase in knowledge of the physical world at this time. The net result was that the first half of the nineteenth century, in particular, was a period of rapid progress along a broad front in mapping, especially in thematic mapping. Commercial and government map publishing increased greatly, leading to the expansion of existing facilities and creation of new ones. Atlases, notably Adolf Stieler's *Hand Atlas* (1817–22 in fifty sheets), and wall maps such as those of Emil von Sydow made the new material available to large numbers of students and to the general public. The uses of globes, maps, and atlases also became important school subjects at this time.

We have discussed the beginnings of modern topographic mapping in France and its extension to other countries. Maps of this type provide a base for data

[1] William Petty, *Political Arithmetick* (London: R. Clavel and Hortlock, 1690).

Diversification and Development in the Nineteenth Century

7

which is most meaningful when expressed at large scale—for example, land use and geology. Broad categories of land use had been shown on early estate and county maps by various hand painted and engraved symbols, and Christopher Packe had distinguished between arable land, downland, and marsh in his map of 1743, which was discussed previously. A more ambitious and systematic attempt in this cartographic genre was Thomas Milne's land use map of London in 1800, drawn at the detailed scale of 2 inches to the mile. Milne used colors, applied by hand, and letters to indicate some seventeen categories of land use, producing a map not unlike those of the Land Utilization Survey of Britain of over a century later.[2] Land use maps show, directly or indirectly, the surface cover—crops, forests, urban forms, etc.—at a given time. These data are ephemeral and in this respect contrast with the more permanant information on geological maps, rock formations with the overburden stripped away.

Although tailing piles give archeological evidence of an interest in minerals before the written record appears, geology, in the modern sense, was founded at the end of the eighteenth and the beginning of the nineteenth century through the efforts of a number of scientists, including James Hutton, Abraham Werner, and Leópold Cuvier. A contemporary of these workers, the English civil engineer William Smith (1769–1839) is credited with first adequately correlating fossils with associated strata and then developing geological mapping.[3] Smith's cartographic efforts were more ambitious and more successful than any previous attempts in this direction. In 1815, after nearly a quarter of a century of research and observation he published his map "The Strata of England," a small part of which is shown as Figure 7.1. This hand-colored map is all the more remarkable when we realize that the data were gathered by noting the strata along canal and road cuts long before the period of the bulldozer. From the geological mapping of William Smith, a conventional, international color and notation scheme for rock types, based on their age and lithology, was subsequently developed. Geologists, including Smith, made important use of profiles or vertical cross sections for showing the relative altitude and the attitude of rocks (Fig. 7.2).

Two of the founders of modern geography, Karl Ritter (1779–1859) and Alexander von Humboldt (1769–1859), were associated with the great cartographic publishing house of Justus Perthes in Gotha. In 1806 Ritter, who was especially interested in geographic education, published

[2] R. A. Skelton, "The Early Map Printer and His Problems," *The Penrose Annual*, vol. 57 (1964), 171–86, esp. 182–84 including color plate.

[3] Kirtley F. Mather and Shirley L. Mason, *A Source Book in Geology*, (New York: McGraw-Hill, 1939), esp. pp. 201–4.

FIG. 7.1. *Portion of William Smith's Geological Map of England (1815).*

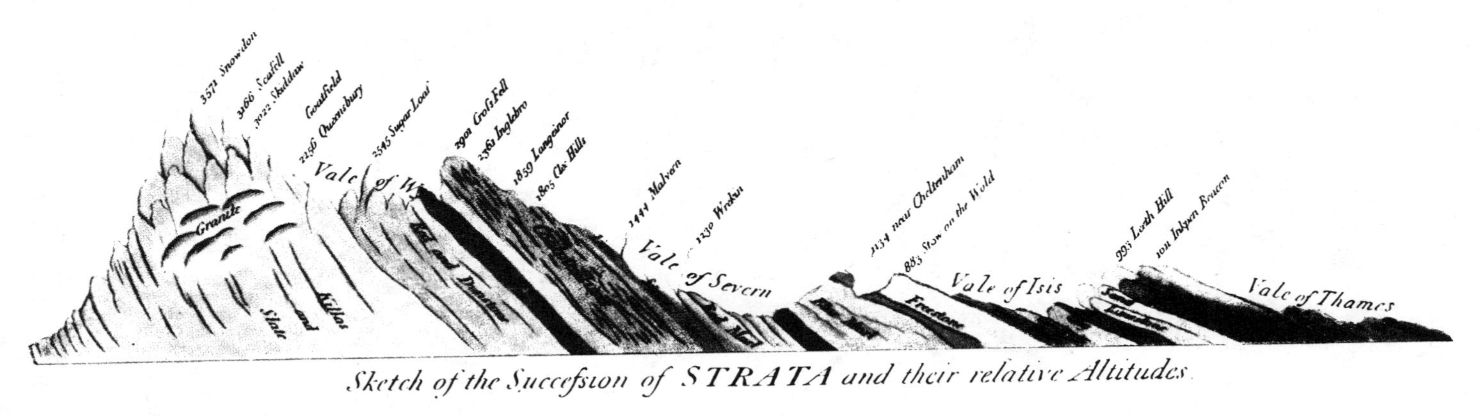

FIG. 7.2. *Profile and cross section of Strata of England, by William Smith (1815).*

a hyposometric map of Europe in which he rendered particular altitude zones by means of bands of gray of decreasing intensity with elevation. Although not the first hypsometric tint map (apparently, as we have indicated, Dupain-Triel may be credited with this development), Ritter's map systematically employs the convention of the higher the altitude, the lighter the tone used. The opposite method, i.e., increase in intensity of tone with elevation, had been employed in a very generalized hypsometric map of the world by A. Zeune, published in 1804.[4]

Humboldt occupied a particularly influential position in the world of science in the first half of the nineteenth century. He was personally acquainted with many of the greatest thinkers of his time and became a statesman of science. Humboldt's early work was in mineralogy, chemistry, and botany, but later he embraced a wide range of knowledge, including physics, oceanography, and climatology in his research. It was in the last named field of study that he made his most original contribution to cartography. Humboldt's isothermal map (Fig. 7.3), published in 1817, resulted from his travels and observations on both sides of the Atlantic.[5] He had noted that the average temperatures on the west sides of continents in the mid-latitudes are milder, by and large, than those on the east coasts in the same latitudes, an idea which overthrew the classical idea of a strict zonality of climate according to latitude. To demonstrate this important concept he drew a plane chart from the equator to 85 degrees N latitude and from 94 degrees W longitude eastward to 122 degrees E longitude, based on the prime meridian of Paris. Within this framework every tenth parallel from 0 to 70 degrees N is drawn, but only three meridians. The average summer and winter temperatures of thirteen places are plotted in their geographical locations but there are no coastlines or other geographical data. To this base Humboldt added isotherms (lines of equal average temperature) which reach their highest latitude at 8 degrees E longitude while the lows are at 80 degrees W and 116 degrees E longitude (the three drawn meridians). The curving isothermal bands contrast with the straight geographical parallels of the plane chart. Below the map Humboldt added a diagram to show the effect of altitude on the isotherms. Humboldt acknowledged his debt to Halley

[4] See Joseph Szaflarski, "A Map of the Tatra Mountains drawn by George Wahlenberg in 1813 as a Prototype of the Contour-line Map," *Geografiska Annaler,* vol. 41 (1959), 74–82.

[5] Arthur H. Robinson and Helen M. Wallis, "Humboldt's Map of Isothermal Lines: A Milestone in Thematic Cartography," *The Cartographic Journal,* vol. 4, no. 2 (1967), 119–23.

FIG. 7.3. (opposite) *Diagrammatic map of isotherms in the Northern Hemisphere, by Alexander von Humboldt (1817).*

Carte des lignes Isothermes par M. A. de Humboldt

Taf. II.

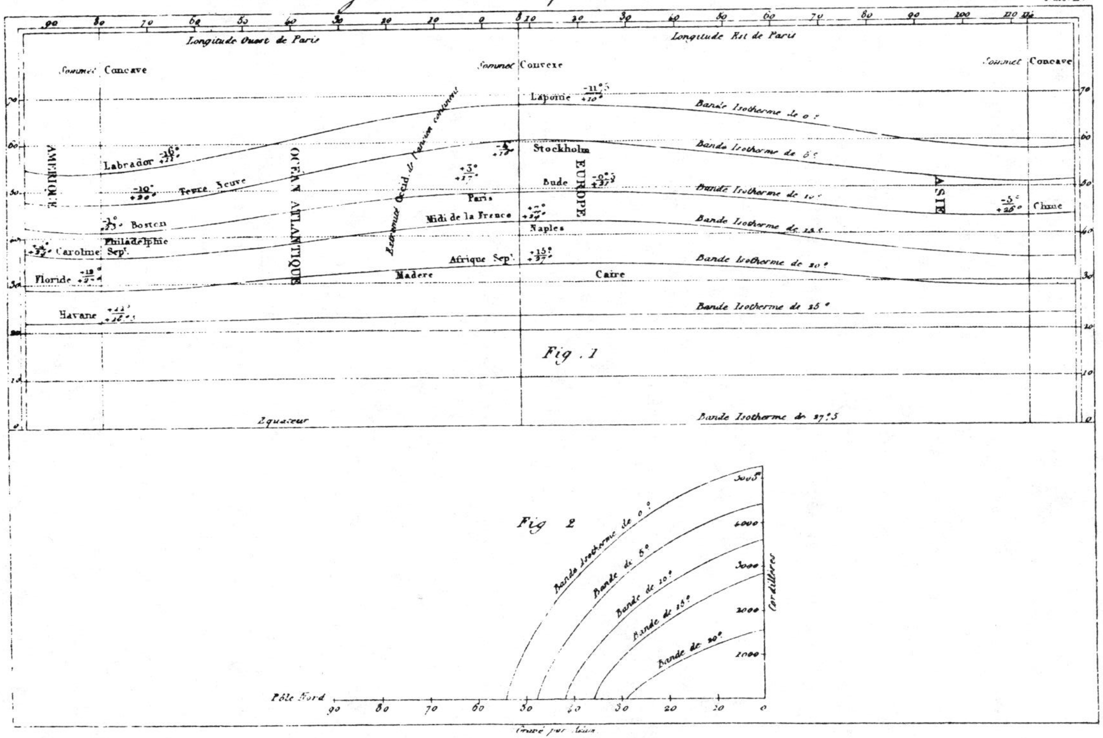

90 80 70 60 50 40 30 20 10 0 10 20 30 40 50 60 70 80 90 100 110

Longitude Ouest de Paris

Longitude Est de Paris

Sommet Concave

Sommet Convexe

Sommet Concave

70

Laponie −11°,5 / +10°

Bande Isotherme de 0°

60

Labrador −16° / +11°

Stockholm

Bande Isotherme de 5°

AMERIQUE

OCEAN ATLANTIQUE

EUROPE

ASIE

50

Terre Neuve −10° / +20°

Paris +3° / +17°

Bude

Bande Isotherme de 10°

Chine −5° / +26°

Boston

Midi de la France

Naples

Bande Isotherme de 15°

40

Philadelphie

Caroline Sept.

Afrique Sept. +15° / +37°

Bande Isotherme de 20°

Floride

Madere

Caire

30

Havane

Bande Isotherme de 25°

20

Fig. 1

10

Equateur

Bande Isotherme de 27°,5

0

Fig 2

Bande Isotherme de 0°

Bande de 5°

Bande de 10°

Bande de 15°

Bande de 20°

5000

4000

3000

2000

1000

Cordillières

Pôle Nord

90 80 70 60 50 40 30 20 10 0

Gravé par Adam

for development of the isoline concept, which was soon applied to other phenomena including early examples in the American atlas of Elias Loomis. Isothermal maps were popularized by Heinrich Berghaus in his *Physikalischer Atlas* (1838–45), which work includes maps that show average barometric pressure at sea level (isobars), average annual precipitation (isohyets), etc. A Dane, Niels F. Ravn, in 1857, used the isoline technique to map population (isodems or isopolanthropes). There are now about one hundred different isolines identified by name, some of the more common of which are listed in Appendix B.

In Chapter 6, we alluded to mapping in America in Colonial times and indicated the strong European influence in this endeavor. Although during that period most maps of America were printed in England, there were, after the middle of the seventeenth century, some notable exceptions—including Lewis Evans' map, "The Middle British Colonies, 1755." This was engraved by James Turner and is believed to have been printed by the press of Benjamin Franklin. Franklin, a Fellow of the Royal Society, drew a chart showing the limits of the Gulf Stream in 1775 based on temperature from Fahrenheit thermometer readings. In spite of these American mapping activities, good maps were not available to the Colonists in the Revolutionary War, as recognized by George Washington who took steps to remedy this situation. Even after the War some American maps continued to be published in England, including Thomas Jefferson's "Map of the Country Between Albemarle Sound and Lake Erie, 1787."[6] But the seeds of a distinctive American school of cartography, in which the military surveyor, as in Europe, played a major role, were sown during the War.[7]

In the previous chapter, we also traced the development of terrain representation in Europe up to the systematization of the hachure by J. G. Lehmann and the use of contours for landforms by J. L. Dupain-Triel. The application of these methods to official American cartography is well illustrated by a map (Fig. 7.4) produced by two early graduates of

[6] Coolie Verner, "Mr. Jefferson Makes a Map," *Imago Mundi,* vol. 14 (1959), 96–108. Of this map, which was based on the Fry-Jefferson map and included in his *Notes on the State of Virginia,* Jefferson declared, the "map was of more value than the book in which it appeared."

[7] Herman R. Friis, "A Brief Review of the Development and Status of Geographical and Cartographical Activities of the United States Government: 1776–1818," *Imago Mundi,* vol. 19 (1965), 68–80.

FIG. 7.4. (opposite) *Comparative maps of Salem, Massachusetts, illustrating the contour (left) and hachure (right) methods of terrain representation by George Whistler and William McNeill (1822).*

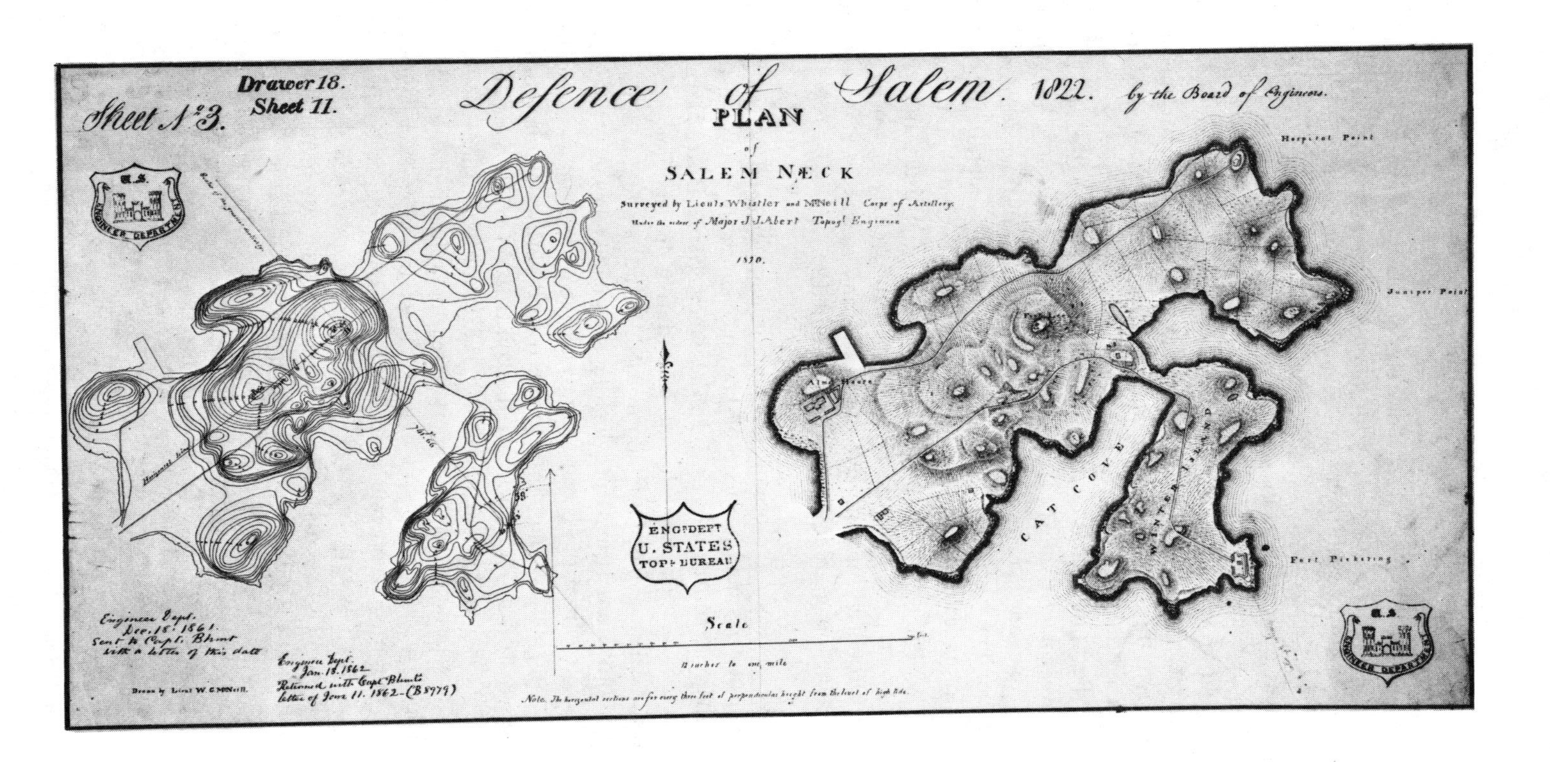

Defence of Salem. 1822. by the Board of Engineers.
Drawer 18.
Sheet 11.
Sheet No 3.
PLAN
of
SALEM NECK
Surveyed by Lieuts Whistler and McNeill Corps of Artillery.
Under the orders of Major J.J. Abert Topog! Engineers
1820.
Hospital Point
Juniper Point
CAT COVE
WINTER ISLAND
Fort Pickering
ENG! DEPT
U. STATES
TOP! BUREAU
Scale
8 inches to one mile
Engineer Dept.
Dec. 18. 1861.
Sent to Capt. Blunt
with a letter of this date
Engineer Dept.
Jan. 18. 1862
Returned with Capt Blunt's
letter of June 11. 1862 - (B8979)
Drawn by Lieut W. G. McNeill.
Note. The horizontal sections are for every three feet of perpendicular height from the level of high tide.

the United States Military Academy, Lieutenants George W. Whistler and William G. McNiell.[8] Whistler was the father of James Abbot McNiell Whistler, the painter, whose own training as a draftsman and map engraver in the Coast and Geodetic Survey was of great value in his career as an etcher. Figure 7.4 illustrates the comparative merits of the contour and the hachure methods. As indicated before, but most clearly shown here, the hachures are short strokes drawn downslope—the thicker the line the steeper the slope. Absolute elevation can only be inferred from the expressive and graphic hachures. By contrast, the contours make it possible not only to read elevation, but also to measure slopes within the limit of the contour interval (i.e., vertical distance between successive contour lines). However, the contour method is difficult for some map readers to understand, being a more abstract form of symbolization. The battle for adoption of the contour (a quantitative method of terrain rendering as opposed to the qualitative hachure) in topographic mapping was not won as early as 1822, the date of Whistler and McNiell's plan of Salem. Actually this method was not officially approved for the British Ordnance Survey maps until 1849 and it was decades after this before most topographic sheets were contoured.[9]

We discussed the origins of thematic mapping and noted the contributions to this field of such well-known scientists as Halley, in the previous chapter, and Humboldt. Like these men, Henry Drury Harness was a thematic cartographer who, at least during his lifetime, was better known in other connections.[10] Born in 1804, Harness had a distinguished career as an administrator for which he eventually received a knighthood. He graduated from the Royal Military Academy at Woolwich and while still a Lieutenant, was employed by the Irish Railway Commission. It was in this service that he supervised the construction of three maps which were published by the Commission in 1837, and which have now established his reputation as an important cartographic innovator. Harness could look to the example of William Playfair (1759–1823) who had devised graphs of economic production and diagrams of relative areas and population by superimposed squares in his *Commercial and Political Atlas* (1786). However, Harness went beyond Playfair in actually mapping statistical data. In his population map of Ireland (Fig. 7.5), Harness used the dasymetric method of showing classes of demographic density

[8] Herman R. Friis, "Highlights of the History of the Use of Conventionalized Symbols and Signs on Large-scale Nautical Charts of the United States Government," in *1st Congrès D'Histoire De L'Océanographie, Bulletin de l'Institute Océanographique* (Monaco 1968), Numero special 2, 223–41.

[9] Szaflarski, "A Map. . . by Wahlenberg," 75.

[10] Arthur H. Robinson, "The 1837 Maps of Henry Drury Harness," *The Geographical Journal,* vol. 71 (1955), 440–50.

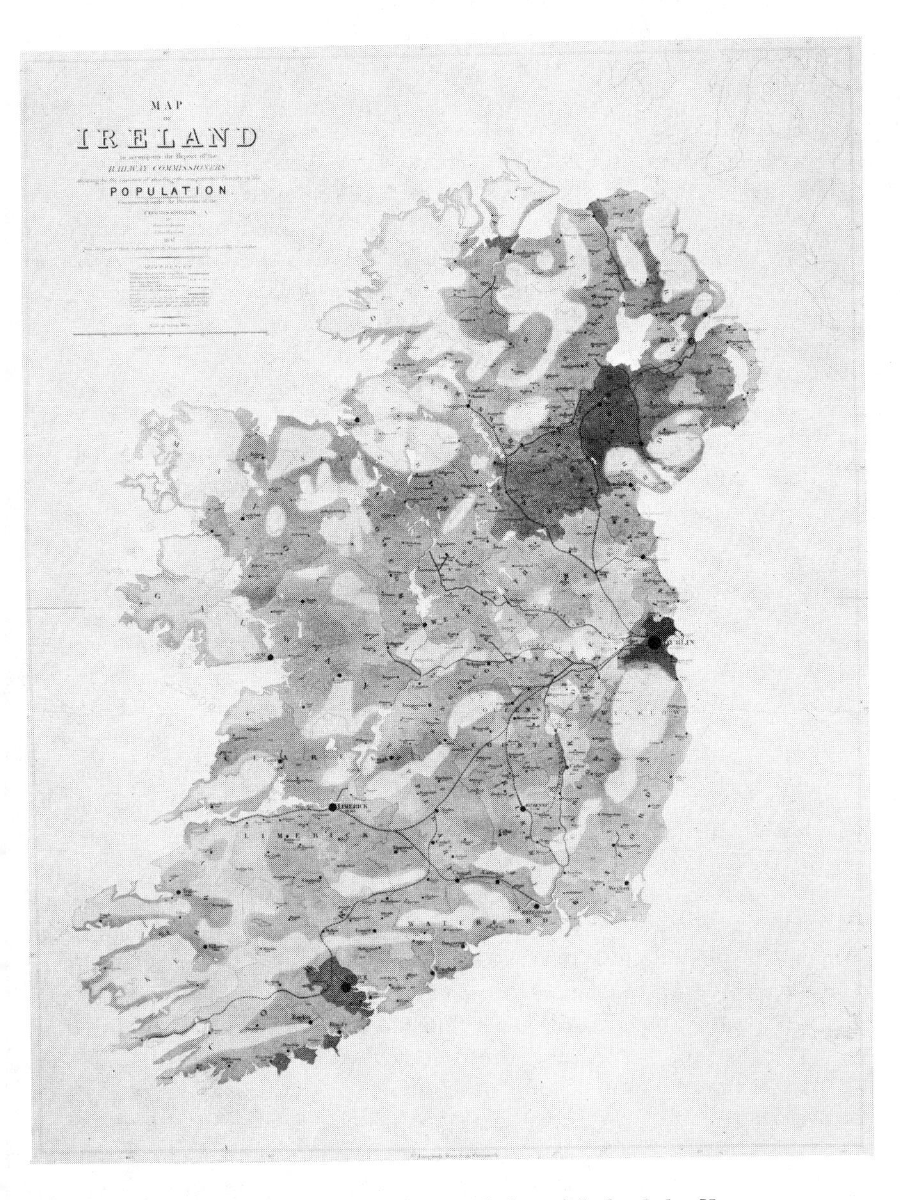

FIG. 7.5. *Dasymetric map of the population of Ireland, by Henry D. Harness (1837).*

in rural areas.[11] In this technique the areal symbols, which in Harness' maps are reproduced by the aquatint process, are not confined by administrative boundaries but cover areas of homogeneity within specified limits. They differ from areas defined by isopleths (such as Humboldt's isotherms) in that higher and lower values can be adjacent to each other without the necessity of intermediate values between. On the same map urban centers are shown by filled circles proportional in size to the population of the place. This same device is used for urban places in Harness' traffic flow map (Fig. 7.6) which also shows flow lines proportional to the relative quantities of traffic in different directions. A third map (not illustrated) employs like symbols to show the number of passengers in different directions by regular public conveyances and, again, the relative sizes of cities are indicated by proportional circles. In his maps for the Railway Commission Harness seems to have originated graduated circles (for cities), density symbols (for population), flow lines (for traffic and conveyances), the dasymetric technique (for population in rural areas), and other carto-statistical methods. August Petermann elaborated and popularized a number of these devices in the middle of the nineteenth century in his widely distributed maps and atlases.

Another carto-statistical technique which seems to have been developed about this time is the dot map which employs symbols of uniform size and value to show the desired distribution. (Uniform symbols for cities do not qualify because what they represent are not of uniform value.) Development of the dot map apparently can be attributed to medical doctors who were concerned with the incidence of cholera at specific locations in the nineteenth century. In 1849 Dr. Thomas Shapter published a map showing deaths from cholera in Exeter, England, for the years 1832–34. For each of the three years he used a different uniform symbol (dot, cross, open circle) and numbers for other associated features. A map which more closely resembles a modern dot map is that of Dr. John Snow (1855) showing deaths from cholera in the Broad Street area of London in September 1854 (Fig. 7.7). The location of each death from the disease, within the period, is marked by a uniform symbol (dark block) and water pumps by crosses. By this means Snow demonstrated, "that the incidence of cholera was only among persons who drank from the Broad Street pump."[12] This map well illustrates the research use of cartography—to find out by mapping that which could not otherwise be learned or, at least, not learned with as great facility and

[11] This method is discussed and illustrated in John K. Wright, "A Method of Mapping Densities of Population with Cape Cod as an Example," *The Geographical Review,* vol. 26 (1936), 103–10.

[12] E. W. Gilbert, "Pioneer Maps of Health and Disease in England," *The Geographical Journal,* vol. 124 (1958), 172–83.

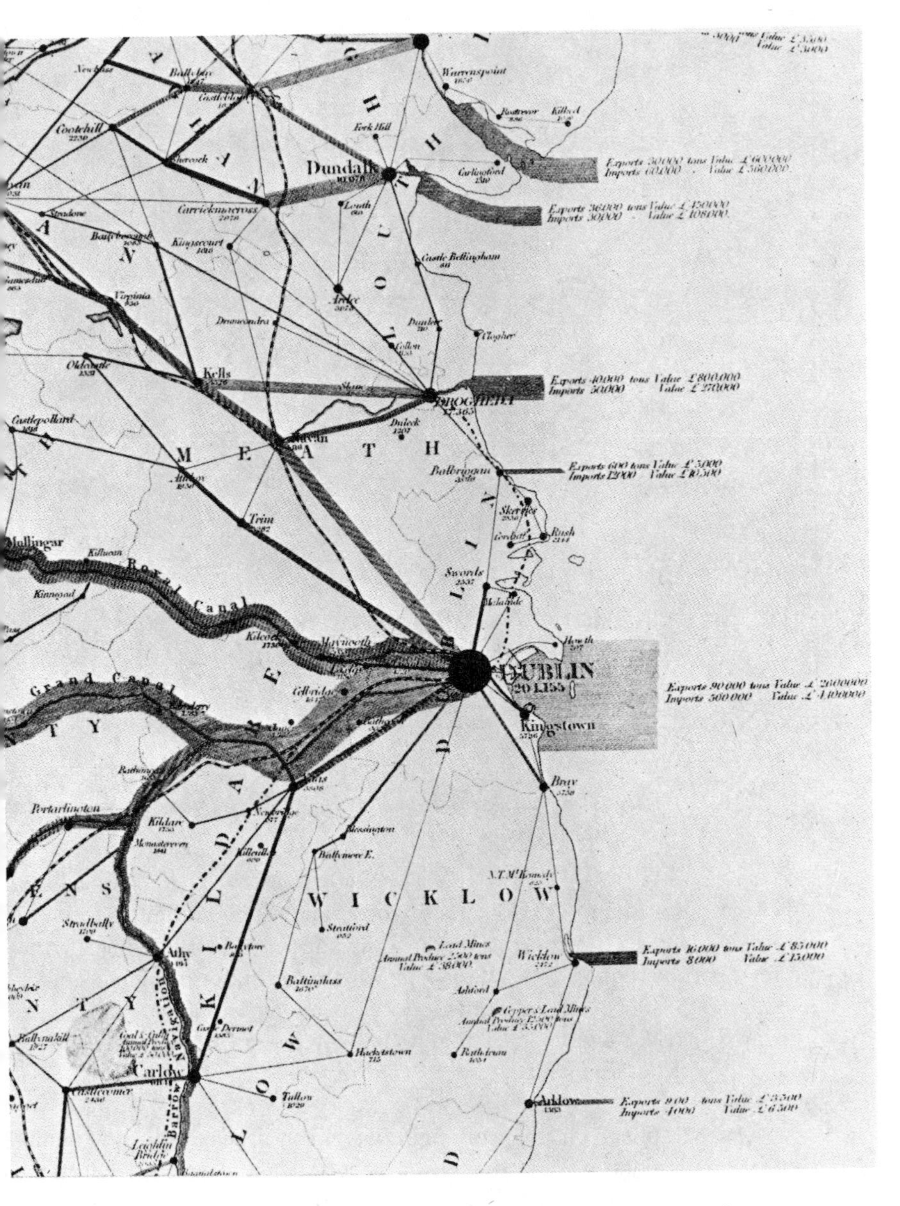

FIG. 7.6. *Portion of traffic flow map of Ireland utilizing quantitative flow lines and graduated circles, by Henry D. Harness (1837).*

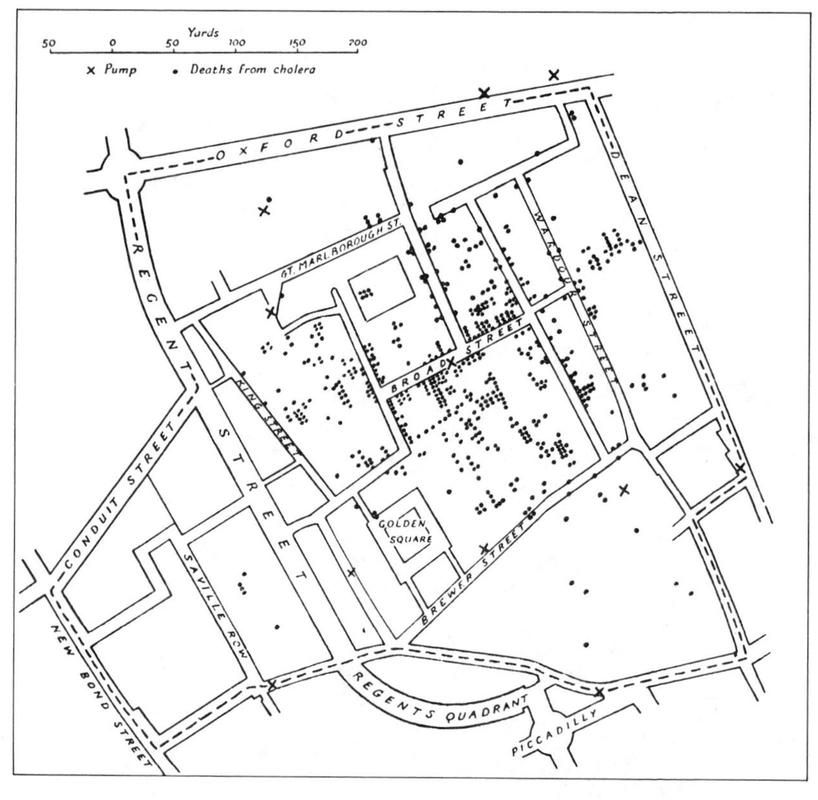

FIG. 7.7. *Dot may (reconstructed) illustrating deaths from cholera in London, by Dr. John Snow (1855).*

precision. The maps of Doctors Shapter and Snow differ in a very important respect from most modern dot maps in that the value of the uniform symbol equals one. A map in which the uniform symbol equals a number greater than one, obviously a higher order of generalization, appeared at least by 1863. In that year a map showing a part of New Zealand with a series of crosses each representing 100 Maoris, was published.[13]

Cartographers on the continent were concerned with the quantitative representation of disease—among them, August Petermann who, as noted previously, worked to extend and popularize statistical cartography. Another direction was the mapping of the distribution of crime, an example of which is A. M. Guerry and A. Balbi's *Statistique Morale de L'Angleterre comparée avec de la France* (1864). In this

13 Raymond P. Hargreaves, "The First Use of the Dot Technique in Cartography," *The Professional Geographer,* vol. 13 (1961), 37–39.

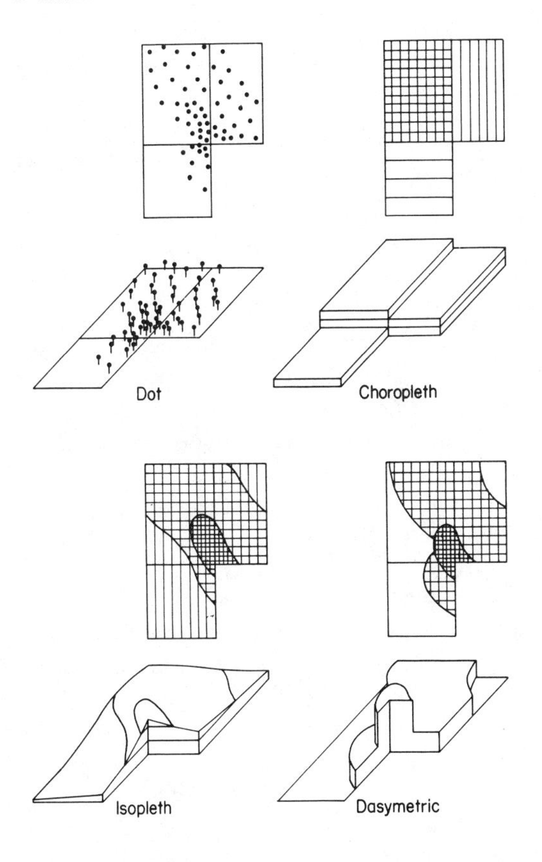

FIG. 7.8. *Diagram illustrating in two- and simulated three-dimensional form principal methods of statistical mapping.*

work we find the simple choropleth technique of quantitative mapping. By this method the amount of the phenomena being mapped is considered in relation to the size of the statistical areas (e.g., counties), density is then computed, and density categories set up. The whole of a particular area is colored (or shaded) uniformly according to the category into which it falls. Obviously in its lack of internal differentiation within a statistical unit the choropleth method differs from the dasymetric, isopleth, and dot techniques (see Fig. 7.8 for graphical amplification of this and other differences between these methods). The flow map, devised by Harness, was popularized by Charles Minard (1781–1870) in his *cartes figuratives.*[14] In his numerous diagrammatic cartographic

[14] Arthur H. Robinson, "The Thematic Maps of Charles Joseph Minard," *Imago Mundi,* vol. 21 (1967), 95–108.

presentations Minard, who was willing to alter spatial relationships to suit his purposes, deals quantitatively with various facets of economic, social, and historical geography. Minard also used the graduated circle and sectored some of these to make so-called "pie graphs"; these have since become popular carto-statistical devices. All of the common techniques of thematic cartography in use today had been developed by 1865.

Although at this period thematic mapping was largely, though not exclusively, a European domain, other branches of cartography were undergoing important development elsewhere. In particular, marine charts of greater sophistication and usefulness to sailors than any produced up to that date were prepared under the direction of Matthew Fontaine Maury (1806–73) in the United States.[15] Maury entered the United States Navy in 1825 and soon turned his attention to a variety of navigational problems. His writings on scientific subjects led to his appointment, in 1842, as Superintendent of the Depot of Charts and Instruments, whose functions were later divided between the United States National Observatory and the Hydrographic Office. It was during his period as Superintendent that Maury produced his celebrated wind and current charts. At first he used old log books as a source of data, but later he requested masters of naval and commercial vessels to send in reports of their voyages on forms prepared specially for this purpose. Generalizing from the great mass of information he received, Maury made recommendations for the quickest passages between various ports according to winds, currents, etc. These tracks, which in some instances deviated considerably from great circle routes, led to savings of days and even weeks on long voyages. A portion of an engraved chart of the Pacific off California by Maury is featured in Figure 7.9; the accompanying legend or key indicates the symbols used on such charts. Maury prepared maps showing various facets of the physical characteristics of the world's oceans but his wind and current charts, as illustrated, are best known. He received international acclaim for his endeavors, being the most decorated American by foreign governments up to his time. Ironically, Maury's recommendations came at a time when sail was soon to be superseded by steam which liberated shipping, to a large extent, from dependence on wind and current. Maury made deep sea soundings, as others had done earlier, but not enough data could be collected by methods then available to permit a satisfactory map of the ocean floor

[15] Matthew Fontaine Maury, *Explanations and Sailing Directions to Accompany the Wind and Current Charts,* 7th ed., (Philadelphia: E. C. and J. Biddle, 1855); Charles Lee Lewis, *Matthew Fontaine Maury: The Pathfinder of the Seas* (Annapolis: The United States Naval Institute, 1927).

to be made. He summarized his ideas in *The Physical Geography of the Seas* (published in 1855) which, along with his other writings and charts, assured him an imperishable place as the founder of systematic oceanography and "the Pathfinder of the Seas."

We have identified a number of directions in which cartography moved in the nineteenth century. The hydrographic offices of the maritime powers carried out detailed surveys of the world's coastlands. In this connection, it is interesting to recall that Charles Darwin was a naturalist on the surveying vessel, H.M.S. *Beagle,* when he made his momentous discoveries leading to the theory of evolution. A second direction was the production of topographic maps of the more heavily settled parts of the land, including various surveys in Europe and the Survey of India. Although some refinements were made in instruments and methods, in general the techniques developed in the previous century were used in these endeavors. In the United States, surveyors built on the work of such men as the Swiss scientist, Ferdinand Hassler (who was appointed Superintendent of the U.S. Coast Survey in 1816) and the Englishman William Tatham, and made serious attempts to standardize the symbols used on maps and charts.[16] A third direction was reconnaissance mapping of the generally sparsely settled interiors of continents—North and South America, Asia, Africa, and Australia. Many geographical discoveries concerning the sources of rivers, the alignment of mountain ranges, the extent of deserts, etc., were recorded on sketch or reconnaissance maps. The names of those who contributed to the unlocking of these secrets and to the more correct cartographic representation of these areas are numerous and well-known, e.g., David Livingstone in Africa, Alexander von Humboldt in South America, and John Wesley Powell in North America.[17] Less spectacular, but not less important, was the cadastral mapping of large tracts of land.

This development is nowhere better exemplified than in maps of the humid lands of North America (although it was repeated, with significant local differences, in similar recently settled areas in other continents). The great Public Domain the young Republic fell heir to after the Revolutionary War in time was expanded to include a vast tract from the Atlantic to the Pacific—the conterminous states of the United States. The subdivision and settlement of this area was a major problem. For these purposes a committee was set up as indicated earlier under the

16 Friis, "A Brief Review," 78, 80.

17 John N. L. Baker, *A History of Geographical Discovery and Exploration* (Boston: Houghton Mifflin, 1931). This work is an excellent one-volume summary of geographical discovery, including the exploration of the interiors of the continents as well as coastal areas; reprinted 1963, Barnes and Noble, New York.

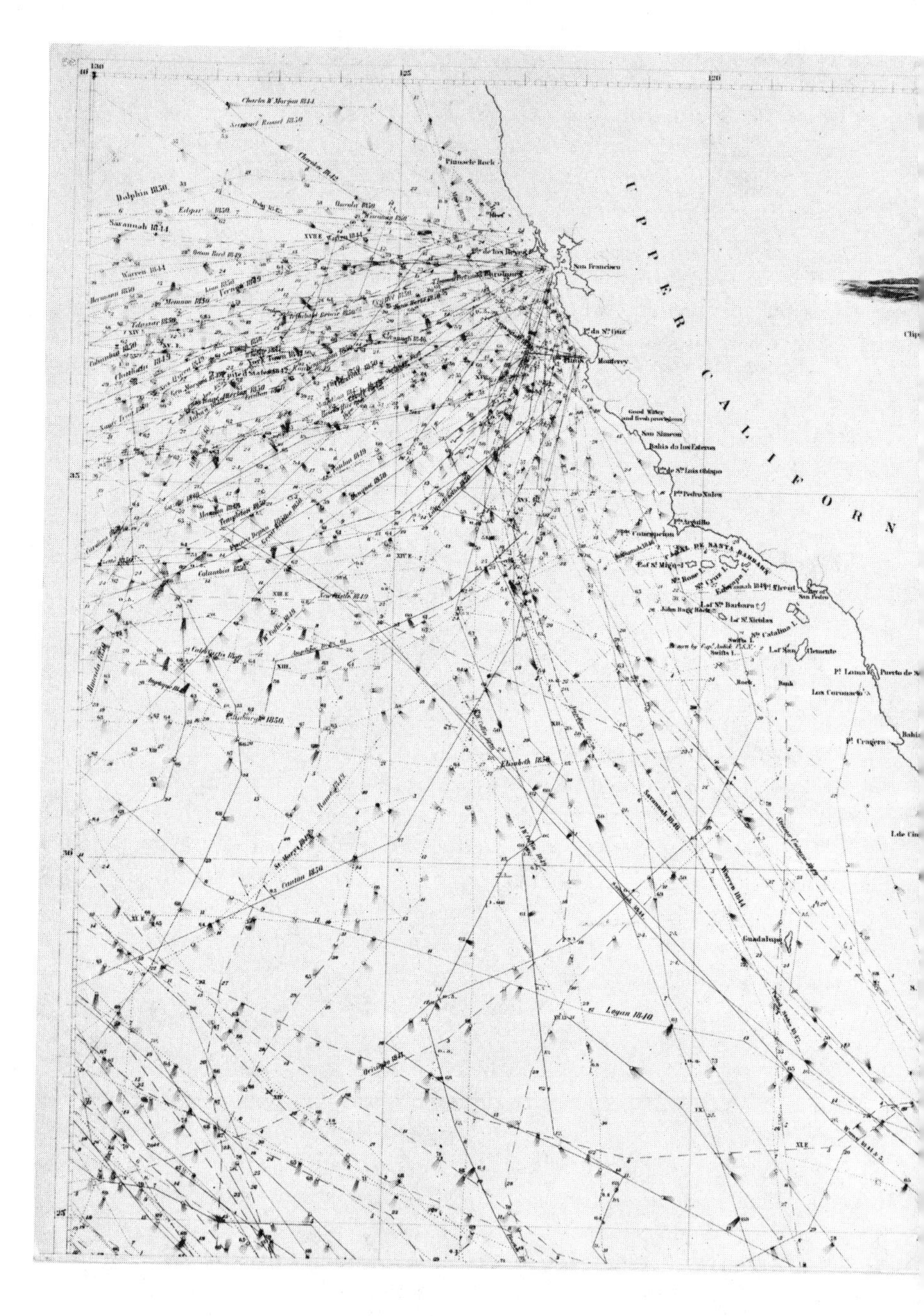

FIG. 7.9. *Chart showing winds and currents off the California coast, by Matthew F. Maury, 1852.*

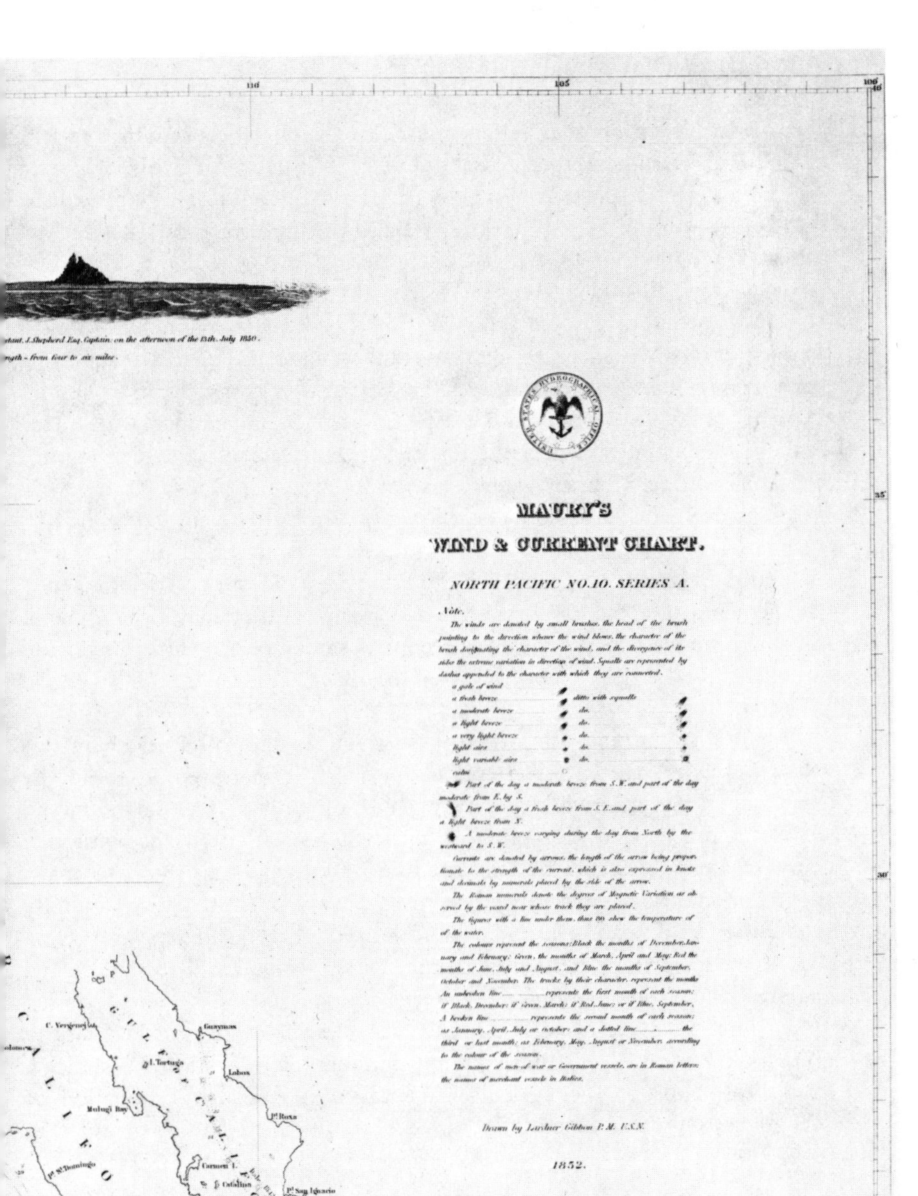
...chant, J. Shepherd Esq. Captain; on the afternoon of the 13th July 1850.
...ngth - from four to six miles.
MAURY'S
WIND & CURRENT CHART.
NORTH PACIFIC NO. 10. SERIES A.
Note.
The winds are denoted by small brushes, the head of the brush pointing to the direction whence the wind blows, the character of the brush designating the character of the wind, and the divergence of its sides the extreme variation in direction of wind. Squalls are represented by dashes appended to the character with which they are connected.
a gale of wind
a fresh breeze ... ditto with squalls
a moderate breeze ... do.
a light breeze ... do.
a very light breeze ... do.
light airs ... do.
light variable airs ... do.
calm
Part of the day a moderate breeze from S.W. and part of the day moderate from E. by S.
Part of the day a fresh breeze from S.E. and part of the day a light breeze from N.
A moderate breeze varying during the day from North by the westward to S.W.
Currents are denoted by arrows, the length of the arrow being proportionate to the strength of the current, which is also expressed in knots and decimals by numerals placed by the side of the arrow.
The Roman numerals denote the degrees of Magnetic Variation as observed by the vessel near whose track they are placed.
The figures with a line under them, thus 80, shew the temperature of the water.
The colours represent the seasons: Black the months of December, January and February; Green, the months of March, April and May; Red the months of June, July and August, and Blue the months of September, October and November. The tracks by their character, represent the months. An unbroken line .... represents the first month of each season; if Black, December; if Green, March; if Red, June; or if Blue, September. A broken line .... represents the second month of each season; as January, April, July or October; and a dotted line .... the third or last month; as February, May, August or November, according to the colour of the season.
The names of men-of-war or Government vessels, are in Roman letters; the names of merchant vessels in Italics.
Drawn by Lardner Gibbon P. M. U.S.N.
1852.
GULF OF CALIFORNIA
CALIFORNIA
C. Virgenes
Guaymas
I. Tortuga
Lobos
Mulugi Bay
Pt. Roxa
Pt. St. Domingo
Carmen I.
Catalina
Pt. San Ignacio
St. Cruz
I. St. Joseph
Pt. St. Lazaro
Espirito Santo
Port Culiacan
M
E
35°
30°
25°

chairmanship of Thomas Jefferson and, as a result of its report, Congress enacted the famous Land Ordinance of 1785. Elements of this legislation which are of particular concern here are: (*1*) survey prior to settlement; (*2*) orientation of survey lines; (*3*) the township unit; and (*4*) the section. To implement this systematic cadastral survey Thomas Hutchins, who had served under George Washington as a surveyor, was appointed Geographer of the United States or first officer in charge of government surveys. Using the magnetic compass and chain as the principal surveying instruments, the United States Public Land Survey was begun in eastern Ohio.[18] As surveys progressed westward some modifications were made in the system but, in general, unsettled land was subdivided into mile square sections. These were organized into townships of thirty-six sections each (6 × 6 sections), blocks of townships being controlled by principal meridians (N-S) and principal bases (E-W). Sections were later subdivided into properties, resulting in a network of fundamental survey lines oriented predominantly in cardinal directions, which stands in sharp contrast to the metes and bounds survey of the eastern seaboard of the United States and of most of the older settled parts of the world (Fig. 7.10).[19]

At first the pace of cadastral surveying in the Public Domain was slow but later, as settlers clamored for land, it increased in speed. By the time of the American Civil War (1861–65) a large part of the humid lands of the eastern half of the United States had been subdivided by surveyors employed by the General Land Office. During the War many of these surveyors enlisted and were employed in a variety of military mapping activities.[20] On the restoration of peace, cadastral surveying again occupied the attention of many workers. In addition to the original survey plats, compiled maps of townships and counties were required for a variety of legal and administrative purposes. The result was a remarkable number of published and unpublished, official and unofficial, cadastral maps of the agricultural areas of America.[21] Production of these maps proved to be a lucrative proposition to publishers who, in

[18] William D. Pattison, *Beginnings of the American Rectangular Land Survey System, 1784–1800,* (University of Chicago, Department of Geography, Research Paper no. 50, 1957).

[19] Francis T. Marschner, *Boundaries and Records in the Territory of Early Settlement from Canada to Florida* (Washington: U.S.D.A. Agricultural Research Service, 1960). A splendid monograph on the problems of unsystematic cadastral surveys particularly, by an author who concerned himself with several aspects of cartography including land use mapping and projections.

[20] A. Philip Muntz, "Union Mapping in the American Civil War," *Imago Mundi,* vol. 17 (1963), 90–94.

[21] Richard W. Stephenson, *Land Ownership Maps: A Checklist of Nineteenth Century United States County Maps in the Library of Congress* (Washington: Library of Congress, 1967). Cartobibliographical studies such as this and

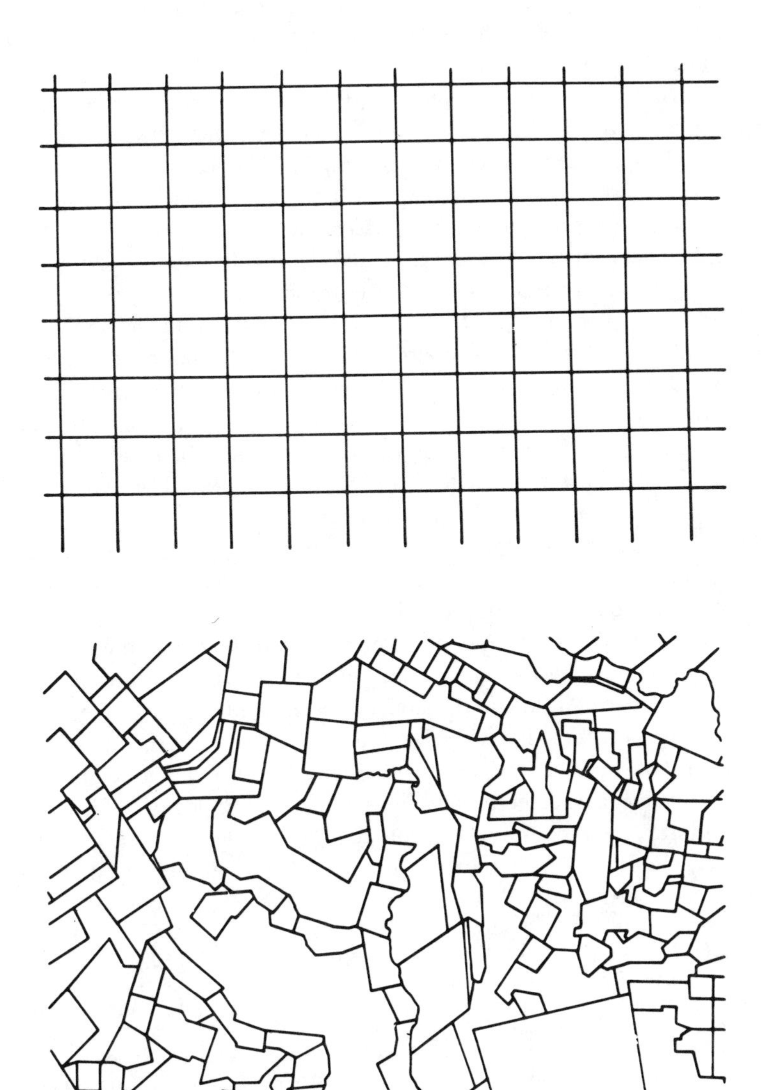

FIG. 7.10. *Map comparing basic cadastral survey units; sections of the U.S. Public Land Survey above, and metes and bounds divisions below. Both embrace one hundred square mile areas in Ohio.*

the 1860's, began to develop county atlases.[22] A phenomenal number of these relatively expensive, hard-covered volumes were printed and sold in the following decades. Before the end of the century some counties in the Corn Belt could boast of ten editions of an atlas, while half of the counties of the United States were not even covered by such atlases. Figure 7.11 is a typical page from a county atlas of the 1870's in the area of systematic surveys. It shows the boundaries of all the properties in a surveyed township, in this case, as frequently, also an administrative township. The property lines and roads commonly extend north-south, east-west without reference to rivers or, for that matter, to the form of the land. As is typical of older cadastral maps, no relief is shown but the size and ownership of properties and the location of dwellings is emphasized. Engraving, and later the cheaper process of wax engraving, was used to reproduce the maps. County atlases were usually sold on a subscription basis and contain pictures of patrons, their families, and the establishments of those willing to pay for this privilege. Characteristically, the pictures are lithographs. The county atlas of the United States well expresses the spirit of the free-enterprise, agricultural society which gave it birth and is a valuable though neglected source of geographical information of the period.[23]

A cartographic form as important for urban places as the cadastral map in rural areas in the nineteenth century, is the fire insurance and underwriters map. Although developed in England in the last years of the eighteenth century, it reached its zenith in the United States in the second half of the last century. These remarkably detailed maps show the fire resistive character of buildings and other types of information. The need for their constant revision led to the development of other means of recording these data in the twentieth century so that they have largely been superseded and are now of historical interest.[24] This

---

the works of Clara Le Gear (footnote 22 following) are of great value to the research scholar. Sometimes the works are topical, as in these cases, or regional as, for example, in the American cartobibliographic studies of William Cumming, Carl Wheat, and Henry Wagner.

[22] Clara E. Le Gear, "United States Atlases, A List of National, State, County, City and Regional Atlases in the Library of Congress" (Washington: Library of Congress, 1950), and by the same author "United States Atlases, A Catalog of National, State, County and Regional Atlases in the Library of Congress and Cooperating Libraries" (Washington: Library of Congress, 1953); and Norman J. W. Thrower, "The County Atlas of the United States," *Surveying and Mapping,* vol. 21, no. 3 (1961), 365–72. Clara Le Gear lists some four thousand different United States county atlases in the two volumes indicated above.

[23] Norman J. W. Thrower, *Original Survey and Land Subdivision: A Comparative Study of the Form and Effect of Contrasting Cadastral Surveys* (Chicago: Association of American Geographers, 1966).

[24] Walter W. Ristow, "United States Fire Insurance and Underwriters Maps: 1852–1968," *The Quarterly Journal of the Library of Congress,* vol. 25, no. 3 (July 1968), p. 194–218.

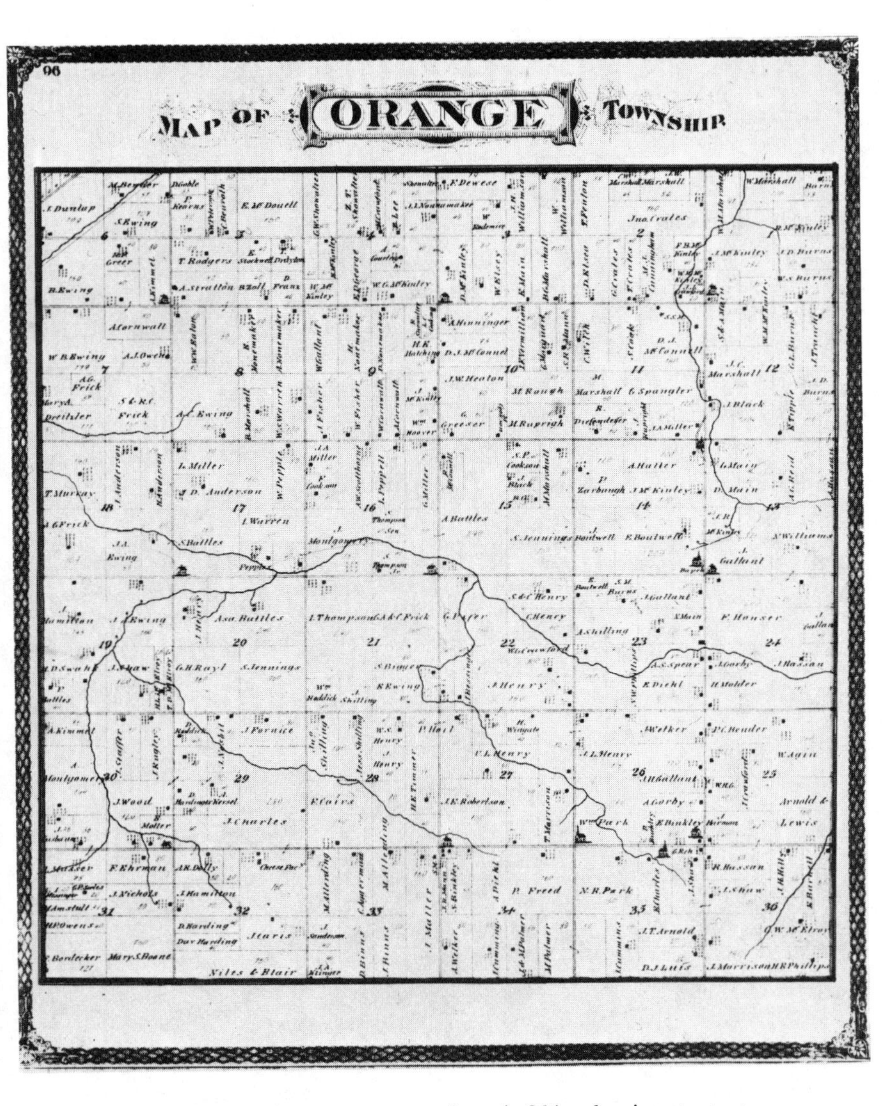

FIG. 7.11. *Page from a county atlas of Ohio showing property boundaries in the 36 sections of a surveyed township; also, in this case, a civil township (1877).*

is one of the relatively few examples of a significant cartographic genre that is now virtually dead.

Another development which found increasing use at this time was the *block diagram.* A block diagram is an oblique or three-quarter view (isometric, perspective, etc.) of a slab of the earth's crust.[25] Surface features are usually portrayed on the upper portion of the block, while on the sides the underlying rock structures are shown. The block diagram thus unites the horizontal dimension of geography with the vertical element of geology, and is a device of particular value in geomorphology—the systematic study of landforms. Geological profiles were prepared by William Smith, as indicated earlier, and Sir Charles Lyell used these devices in rather naturalistic settings, with a suggestion of the associated surface features. Block diagrams were made by mining engineers to show mines, quarries, and caves, especially in Europe. In the hands of American geomorphologists Grove K. Gilbert and William M. Davis, the block diagram with its conventional rock symbols on the sides of the slab was fully developed (Fig. 7.12). If well drawn, the block diagram can be

[25] Armin K. Lobeck, *Block Diagrams and Other Graphic Methods Used in Geology and Geography,* 2nd. ed. (Amherst: Emerson-Trussel, 1958); Norman J. W. Thrower, "Block Diagrams and Mediterranean Coastlands: A Study of the Block Diagram as a Technique for Illustrating the Progressive Development of Low to Moderately Sloping Coastlands of the Mediterranean Region," unpublished B. A. thesis (University of Virginia, 1953).

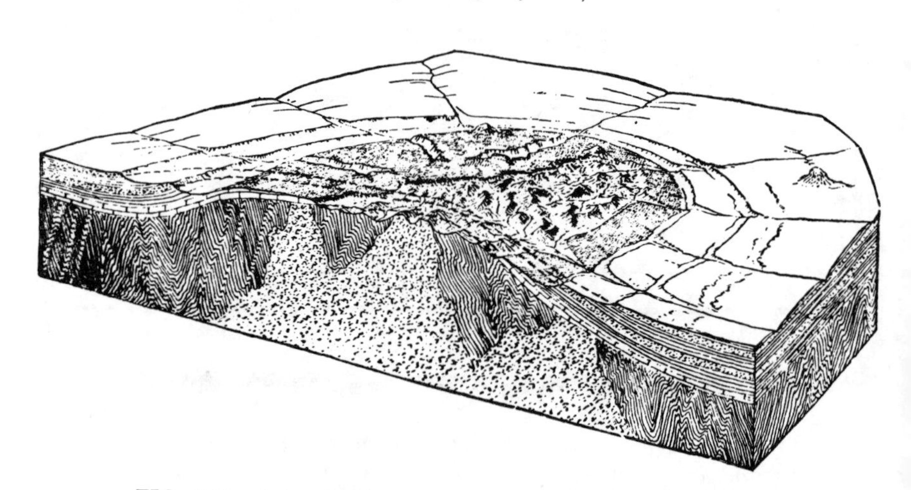

FIG. 7.12. (above) *Block diagram by William Morris Davis illustrating the relationship between surface features and subsurface structures.*

FIG. 7.13. (opposite) *Shaded relief lithographic map of California. Made under the direction of J. D. Whitney, State Geological Survey of California (1874).*

a perfectly consistent view of part of the earth. Davis, who had a fine understanding of graphical means of expression, used series of block diagrams to illustrate temporal changes in the landscape.

Lithography, which was developed by Aloys C. Senefelder (1771–1834) in the last years of the eighteenth century, was first used for maps in the early nineteenth century. This process allows the production of continuous tonal variation or shading which is of great utility in cartography. It was used for some multiple color and shaded relief maps in the nineteenth century, one example being a map of California published with the approval of the University of California in 1874, some five years after the establishment of that institution (Fig. 7.13). This is an

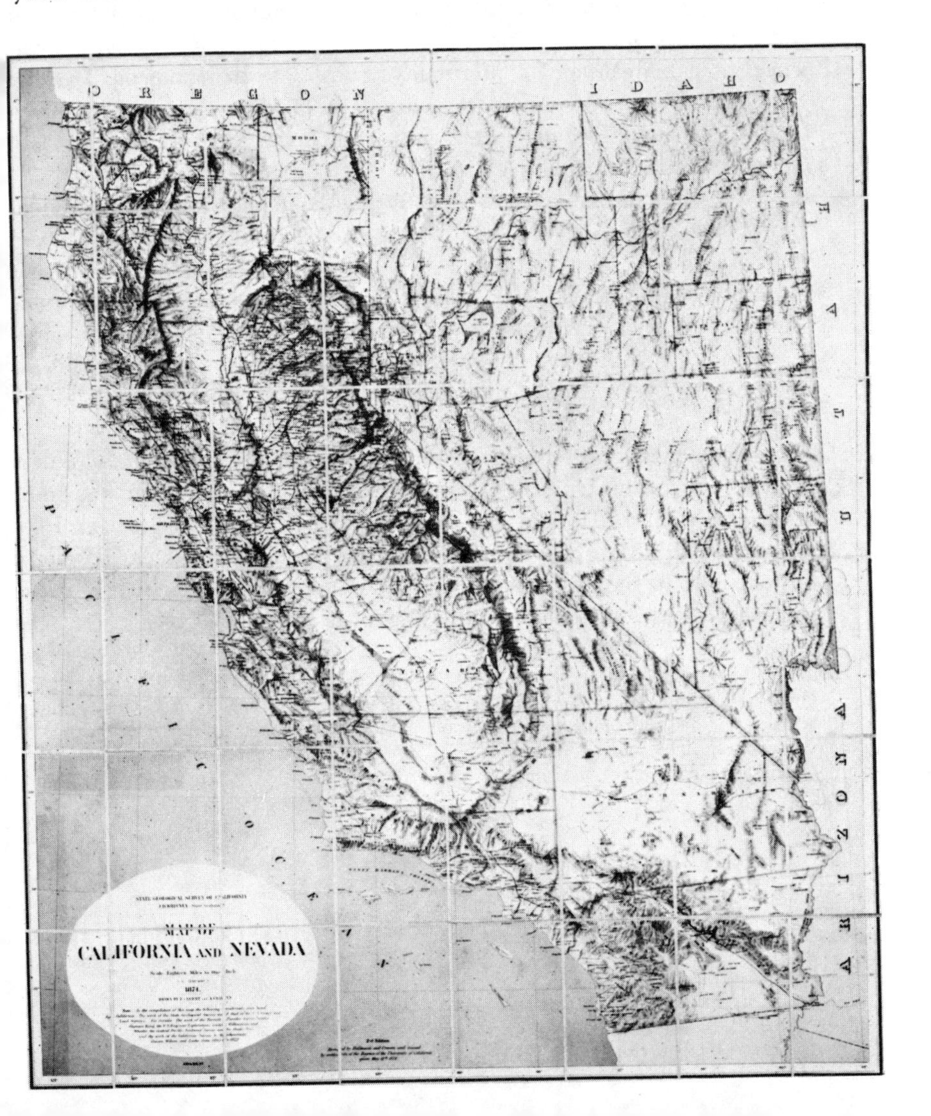

early example of shaded relief by the use of lithography. As we shall see, this became a very important technique in cartography in the twentieth century, particularly after lithography was married to photography. Photographic processes suitable for map reproduction were developed in the second half of the nineteenth century but photography was also soon to provide an improved data source—aerial photography (at first using the balloon as a platform and later the airplane), which has recently revolutionized mapping to as great an extent as did printing in the fifteenth century. [These potentials were not realized until the present century (the engraved map continued to hold its own in the nineteenth century) and so will be considered in our discussion of modern cartography.] New projections were also developed in the nineteenth century, including some with prime qualities, but more had no special characteristics and simply represent a good compromise between equivalence (equal area) and conformality. Such a grid arrangement is the Polyconic projection devised by Ferdinand Hassler. This was to have, with some modifications, special importance in the next century (see Chapter 8 and Appendix A).

We shall have occasion to refer in the next chapter to national atlases, the earliest of which has been revised and is available today—that of Finland, first published in 1899. This was sponsored by the Finnish Geographical Society, one of a number of such organizations founded in various countries in the nineteenth century which did much to promote cartography as well as geography.[26] Actually the *Statistical Atlas of the United States* compiled by Francis A. Walker pre-dates the *Atlas of Finland* by twenty-five years. Walker's volume employs interesting symbolization and color to show a variety of physical and cultural distributions in the United States based on the 1870 census. It was compiled and published under the authority of the U.S. Congress in 1874 but, unlike the *Atlas of Finland,* did not go through subsequent revisions. In fact only in recent times, about 100 years after Walker's work, has a general United States national atlas again been issued.

[26] John K. Wright, "The Field of the Geographical Society," chap. 23 in Griffith Taylor, ed., *Geography in the Twentieth Century* (London: Methuen, 1937), pp. 543–65. This general essay includes references to the cartography of the geographical societies and lists in a footnote individual histories of a number of these societies.

Modern Cartography: Official and Quasi-Official Maps

8

In a sense the modern period of cartography can be said to begin with a formal proposal, in 1891, for an International Map of the World (IMW) at the scale of 1:1,000,000 (0.1 millimeter on the map equals 100 meters on the ground). This proposal was made by Professor Albrecht Penck, a German geographer, at the Fifth International Geographical Congress held in Berne, Switzerland. Some years before this Penck had worked out preliminary specifications for the projected map in the form of a series of standard sheets bounded by parallels and meridians and utilizing uniform symbols. Considerable opposition to this idea had to be overcome but at subsequent congresses the concept gained increasing acceptance. Experimental sheets were produced in several European countries, and in 1909 the First International Map Committee was convened in London to consider general specifications and production methods. Agreement on these was reached at a second meeting of the Committee held in Paris in 1913, which also set up a Central Office at Southampton, the home of Britain's Ordnance Survey. Understandably little work was accomplished during World War I, but after 1921 the Central Office published reports annually. Considerable progress was made in the production of the map sheets between the World Wars but at the end of this period, great areas still lacked coverage—including much of interior and eastern Asia and North America. A remarkable contribution to the series was the "Map of Hispanic America on the Millionth Scale" consisting of over 100 sheets (or about one-tenth of the total in the series) covering South and Middle America, by the American Geographical Society of New York.

Although some IMW sheets were published during World War II, owing to the great increase of flying activity at this time, the need for aeronautical charts became of paramount and immediate concern. To meet this need a new international map of the world on the scale of 1:1,000,000, the World Aeronautical Chart (WAC), was compiled, generally according to a form drawn up by the International Civil Aviation Organization. Most of the work was accomplished in the United States under the direction of the U.S. Aeronautical Chart Service [later the Aeronautical Chart and Information Center and now the Aerospace Center, Defense Mapping Agency (D.M.A.)].

In 1953 the functions of the IMW Central Office were transferred to the Department of Economic and Social Affairs (Cartographic Section) of the United Nations in New York. Since this time a major concern has been the possible duplication of the function of the IMW by WAC. However, it has now been decided to keep the two world map series separate because their objectives are different, the IMW being for general and scientific purposes, and the WAC sheets specifically for use of the aviator. At the same time it was agreed that source materials should be available for the compilation of both series and that consideration should be given to use of a common projection and uniform sheet limits.[1]

Before discussing the nature of the cartography of these two most important international world maps of intermediate scale, it is necessary to understand the political climate and technological progress which made these projects possible. We have noticed the secrecy which often surrounded map-making activities in the past. On the other hand we have also alluded to a "brain drain" of scientists (including cartographers) from one country to another and, later, the connecting of triangulation nets across political boundaries. In spite of the international character of these operations, from the fifteenth to the end of the nineteenth century, mapping was largely for the benefit of individual states. It is a tragic irony that at a time when Europe was emerging from its nationistic parochialism, the continent should be devastated by two World Wars. Nevertheless, through all of the vicissitudes of the present century, the IMW has been a vehicle for international cooperation, if not an unqualified cartographic success.

[1] Federal Republic of Germany, *United Nations Technical Conference on the International Map of the World on the Millionth Scale* (Bonn: Institut für Angewandte Geödasie, 1962); Hans-Peter Kosack, "Cartographic Problems of Representing the Polar Areas on the International Map of the World on the Scale 1:1,000,000 and on Related Series," a background paper for the United Nations, *Technical Conference on the International Map of the World on the Millionth Scale* (Bonn: 1962). The United Nations also sponsors thematic mapping, e.g., UNESCO's "Bioclimatic Map of the Mediterranean Region" by Henri Gaussen and others. Gaussen is a leader in the field of vegetation mapping.

The technology which produced the IMW and, later, WAC projects relates to the progress of topographic mapping, the advance of geodesy, and the development of aviation. Sheets of both international series are compiled maps which rely on information from more detailed sources, especially topographic maps. Topographic mapping, in turn, depends on the accuracy of the triangulation network and the spheroid to which it is related. The Clarke Spheroid of 1866, a theoretical figure, was largely superseded by the International (Hayford) Spheroid adopted by the International Geodetic Association in 1924. Recently satellites have provided data on the figure or shape of the earth which permit more accurate mapping to be undertaken. Triangulation networks have been connected between continents, sometimes spanning great distances over water bodies. For example, the Scandinavian system of triangulation has been connected via the Faroe Islands, Iceland, Greenland, and Newfoundland, to mainland Canada. This was accomplished by the Hiran method whereby a measurement is made of the time taken for an electromagnetic impulse to travel from one point to another. Thus the Old World and New World nets have been connected. We shall consider the impact of the airplane on topographic mapping in a later section of this chapter.

Figure 8.1 shows an IMW sheet greatly reduced in size. Like all maps in the series it has a unique designation, in this case, "N.I 11, Los Angeles." It covers 4 degrees of latitude and 6 degrees of longitude, as do all of the other IMW sheets except those poleward of 60 degrees latitude which, because of the rapid convergence of meridians, cover 12 degrees longitude or more. The projection used is the Polyconic, modified so that adjacent sheets fit on all four sides. The Modified Polyconic, like its parent projection, is neither conformal (orthomorphic) nor equivalent (equal area) but is suitable to the purpose and scale of the map. The hyposometric system of relief is used with principal contours at 200, 500, 1000, 2000, 2500, 3000 meters (up by 1000's of meters) above sea level, the tints between progressing from green in the lowlands through ochre to brown for the mountains. The entire system works well on the Los Angeles sheet, where the option to add auxilliary contours at every 100 meters has been exercised to show relief on the land from below sea level to 3,500 meters. It is not as satisfactory in regions of lower relief differences and less varied landforms—in fact, a major problem of the IMW series has been the creation of symbols that are equally applicable to all parts of the world. Undersea features are delineated by isobaths whose numerical values are the same as those used for dry land contours; the datum for these is mean sea level, and tints of blue are used between selected isobaths progressively deeper in hue the deeper the water.

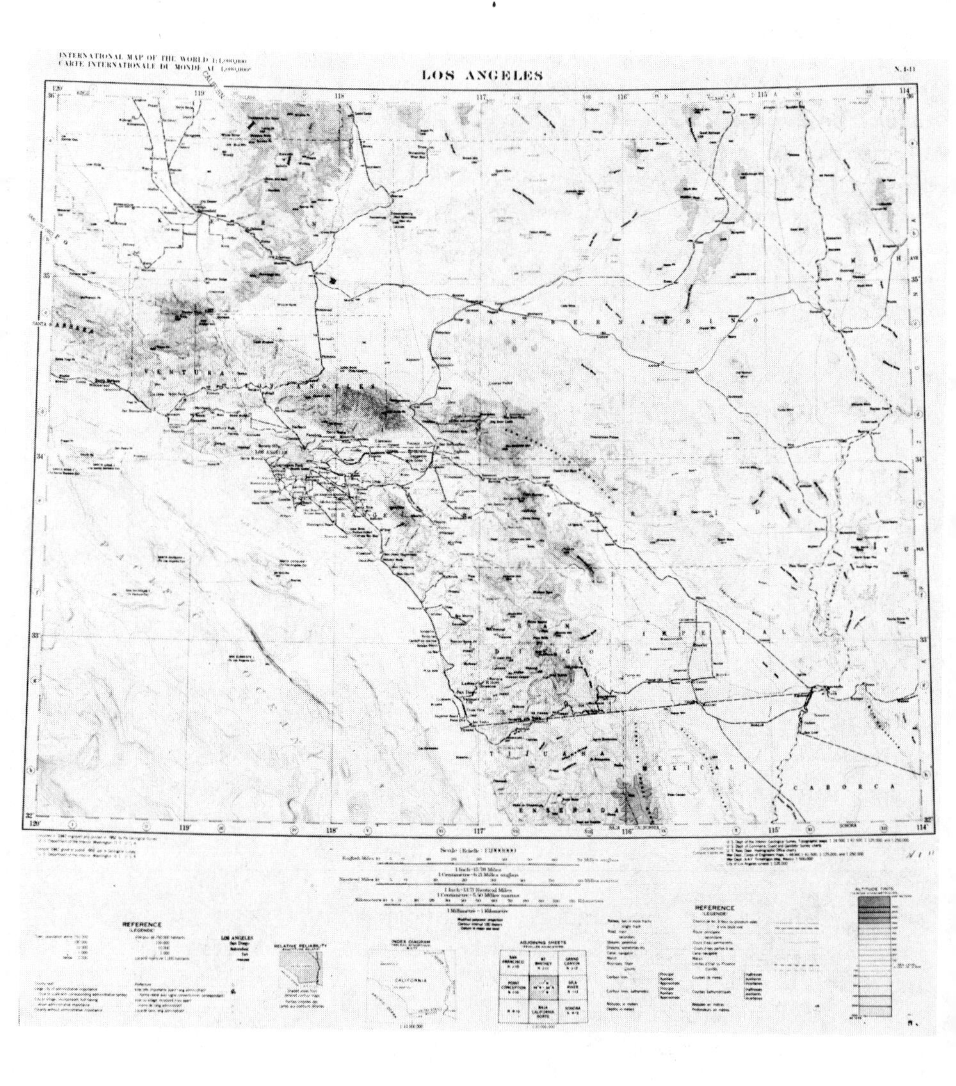
INTERNATIONAL MAP OF THE WORLD 1:1,000,000
CARTE INTERNATIONALE DU MONDE AU 1,000,000
LOS ANGELES
SAN BERNARDINO
VENTURA
LOS ANGELES
SAN DIEGO
IMPERIAL
TIJUANA
MEXICALI
ENSENADA
CABORCA
REFERENCE
REFERENCE
INDEX DIAGRAM
CALIFORNIA
ADJOINING SHEETS
RELATIVE RELIABILITY

Lettering, boundaries, transportation features, etc., conform to the specifications contained in the Resolutions of the IMW of 1909 and 1913.[2]

Opposition to the series has again developed in recent years. Some critics consider the scale not really useful for the intended purposes, and others feel that the symbolization, which reflects the technology of the early years of this century, is badly in need of revision (Fig. 8.2).[3] When the IMW was initiated, it was envisaged as a general planning map which would serve as a base for maps of other distributions—population, ethnic groups, archeology, vegetation, soils, and geology. While examples of maps of all of these phenomena on the IMW base have been produced, generally they are of quite limited extent. The most ambitious effort in this direction appears to be the geological map of the U.S.S.R. which, according to Soviet statements, is to cover the entire country. To some extent the functions of the IMW in the U.S.S.R. and Eastern Europe is satisfied by the *Carta Mira* (World Map) on the scale of 1:2,500,000 in 244 sheets. This general world map is published by the U.S.S.R. and politically affiliated states, each of the co-operating countries having special areas of responsibility.

The IMW on the millionth scale is still unfinished even though revised sheets of some areas previously mapped in this series have already appeared. By contrast, the WAC, on the scale of 1:1,000,000, is complete in that sheets have been made of all areas of the earth.[4] Superficially the appearance of IMW and WAC sheets are similar but there are significant differences between the two series. The Lambert Conformal Conic projection, used for the WAC, has small scale error and relatively straight azimuths over the extent of a single chart and, of course, shapes are correct around a point. Like the IMW, WAC sheets cover 4 degrees of latitude but there is a greater variety in the longitudinal spread on the WAC quadrangles. An interesting feature of the WAC maps is that the back of some of the sheets is used for an index map to the series, for the legend, etc. During World War II, some sheets were printed on cloth to render them less susceptible to moisture damage. A wide variety of

[2] United Nations, "First Progress Report on the International Map of the World on the Millionth Scale (1954)," *World Cartography,* vol. 4 (1954).

[3] Richard A. Gardiner, "A Re-Appraisal of the International Map of the World (IMW) on the Millionth Scale," *International Yearbook of Cartography,* vol. 1 (1961), 31–49.

[4] International Civil Aviation Organization, *Aeronautical Information Provided by States,* 22nd ed. (Quebec: International Civil Aviation Organization, December 1968).

FIG. 8. 1. (opposite) *International Map of the World (IMW) of Los Angeles, California. Original at 1:1,000,000 (greatly reduced).*

INTERNATIONAL MAP OF THE WORLD

SCALE 1:1.000.000

PROPOSED CHANGES TO CONVENTIONAL SIGNS

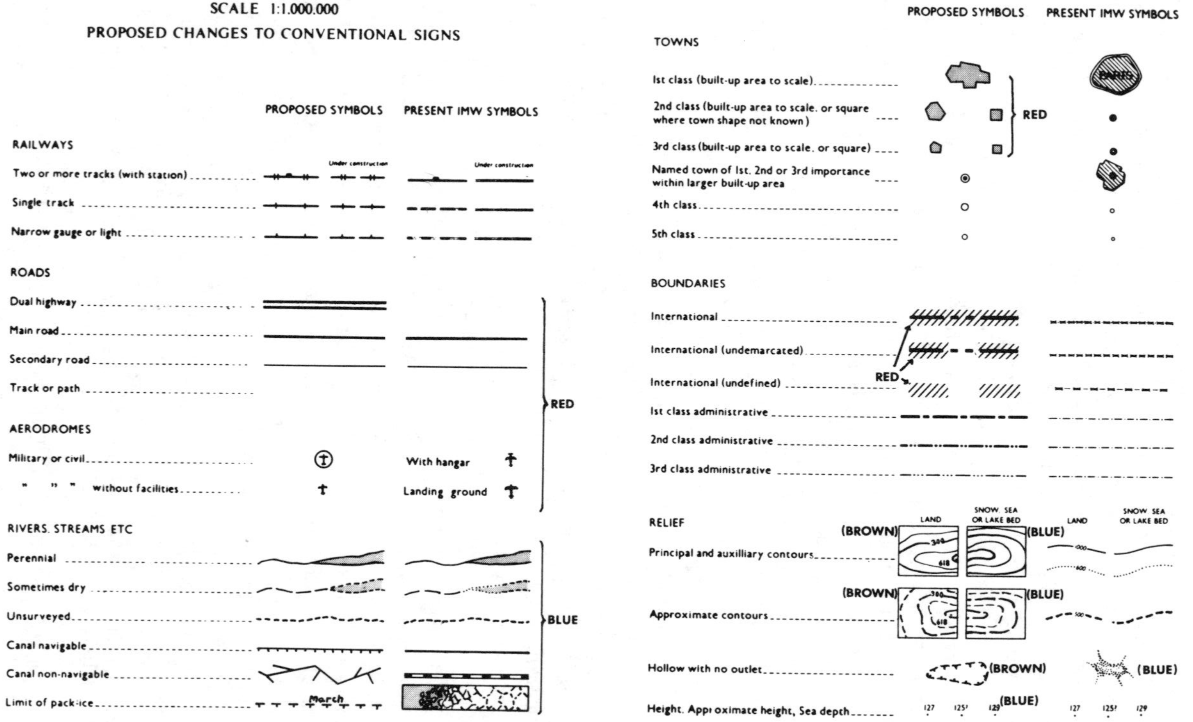

PROPOSED SYMBOLS | PRESENT IMW SYMBOLS

RAILWAYS

Two or more tracks (with station) — Under construction — Under construction

Single track

Narrow gauge or light

ROADS

Dual highway

Main road

Secondary road

Track or path

AERODROMES

Military or civil — With hangar

" " " without facilities — Landing ground

RED

RIVERS. STREAMS ETC

Perennial

Sometimes dry

Unsurveyed

Canal navigable

Canal non-navigable

Limit of pack-ice — March

BLUE

PROPOSED SYMBOLS | PRESENT IMW SYMBOLS

TOWNS

Ist class (built-up area to scale)

2nd class (built-up area to scale, or square where town shape not known)

3rd class (built-up area to scale, or square)

RED

Named town of Ist, 2nd or 3rd importance within larger built-up area

4th class

5th class

BOUNDARIES

International

International (undemarcated)

International (undefined)

RED

Ist class administrative

2nd class administrative

3rd class administrative

RELIEF

LAND | SNOW SEA OR LAKE BED | LAND | SNOW SEA OR LAKE BED

Principal and auxilliary contours — (BROWN) 300 618 — (BLUE)

Approximate contours — (BROWN) 300 618 — (BLUE)

Hollow with no outlet — (BROWN) — (BLUE)

Height, Approximate height, Sea depth — 127 125? 129 (BLUE) — 127 125? 129

information of value to the aerial navigator is contained on the WAC,. which has hypsometric relief similar to that of the IMW. Outside of the United States the Operational Navigation Chart (ONC) shaded relief series is now replacing the WAC series. On all of these charts the quality of the base information varies enormously from area to area depending on the nature of the source materials available for the compilation. Aeronautical charts of smaller scales such as 1:2,000,000 for jet navigation and route planning, as well as those of large scale, e.g., 1:500,000 and 1:250,000 sectional and local charts and even more detailed instrument approach and procedure charts, are available for specific areas. Some of these now use shaded relief for landforms and are printed in colors suitable to the illumination in the cockpit. Thus the airplane has internationalized cartography in very important and positive ways.

That age-old device, the hydrographic chart, has recently undergone important changes as a result of the availability of more and better data. Echo sounding, for example, through the use of automatic depth-finding mechanisms, and new instruments and techniques for determining geographical position at sea, are providing an abundant source of information. Echo sounding or sonar permits continuous depth traces across the ocean floor to be made by a ship in progress so that, at last, the true three-dimensional form of the oceans can be charted. Symbolization has been improved through the use of color and modern printing techniques. The British Admiralty and the U.S. Hydrographic Center, D.M.A., in particular, have produced charts of all oceans. An attempt to coordinate the activities, including charting, of the various hydrographic survey organizations is being made through the International Hydrographic Bureau with headquarters in Monaco. As the resources of the oceans or "inner space" as they have been called are increasingly utilized, even better maps of this greater part of the earth's surface will be needed and will be produced.

In earlier chapters we have traced the development of the large scale topographic map. We have also alluded to the use of this general type of map as a base for other cartographic works of large scale and as the source for compiled maps of intermediate and small scale. The methods by which topographic maps are made have been revolutionized in this century through the airplane and the development of aerial photography. This has given rise to a new branch of engineering—photo-

FIG. 8.2. (opposite) *Existing and proposed symbols for the International Map of the World (reduced).*

grammetry, which is the science of obtaining reliable measurements by means of photography and, by extension, mapping from photos.

In keeping with our intention to emphasize maps rather than mapping, we will confine our remarks concerning photogrammetry to a bare minimum. However, we should re-emphasize that the use of air photographs in the twentieth century has wrought changes in cartography perhaps only comparable to the effect of printing on map-making in the Renaissance. Photogrammetric methods have reduced the cost of making maps remarkably, made it possible to map areas that would otherwise be difficult to reach and, most important, increased the quality and accuracy of mapping generally. It is only necessary to compare topographic quadrangles of the same area surveyed by ground and aerial methods to appreciate this point (Fig. 8.3). The preferred source materials for air surveys are vertical photographs with a sixty percent overlap in the direction of the line of flight of the aircraft taking the pictures and a twenty-five percent overlap with adjacent flights. The overlap ensures that only the most accurate part of any particular photo, the center, need be used and, more significantly, permits stereoscopic analysis. This is accomplished with the aid of the stereoscope and much more elaborate and refined optical instruments (e.g. Multiplex) based on the same principle which makes it possible for a viewer to observe a miniature, three-dimensional model of a given area. It is such models, made by fusing photographic images of the area taken from different places (stations), utilizing the parallax factor, that are used for contouring as well as for planimetric mapping.[5] Because by this method the stereoscopic model, with all its rich detail, is in the photogrammetrist's view, it is obvious why mapping, especially in areas of high relief, is generally superior by aerial than field survey where one must interpolate between a finite number of points of known value. Of course, some horizontal and vertical control points to which the air photo coverage is tied must be determined in the field; first, second, and third order points (the numbers refer to the degree of accuracy of these control points, which differs in different surveys) used for this purpose are indicated on the ground by metal tablets. Field checking, after the map is compiled by photogrammetric means and before it is finally rendered, is most desirable.

It might properly be asked at this point why the vertical air photos are not used in place of topographic maps. The answer is simply that the photos, even though distortions can be removed to make them more or less correct in scale, possess too much information: a map represents a judicious selection of data for particular purposes. Actually controlled,

[5] George S. Whitmore, Morrie M. Thompson, and Julius L. Speert, "Modern Instruments for Surveying and Mapping," *Science,* vol. 130 (1959), 1059–66.

FIG. 8.3. *Layout showing the relationship between an air photo and a map, and maps of different scale and date, etc. Prepared by the United States Geological Survey.*

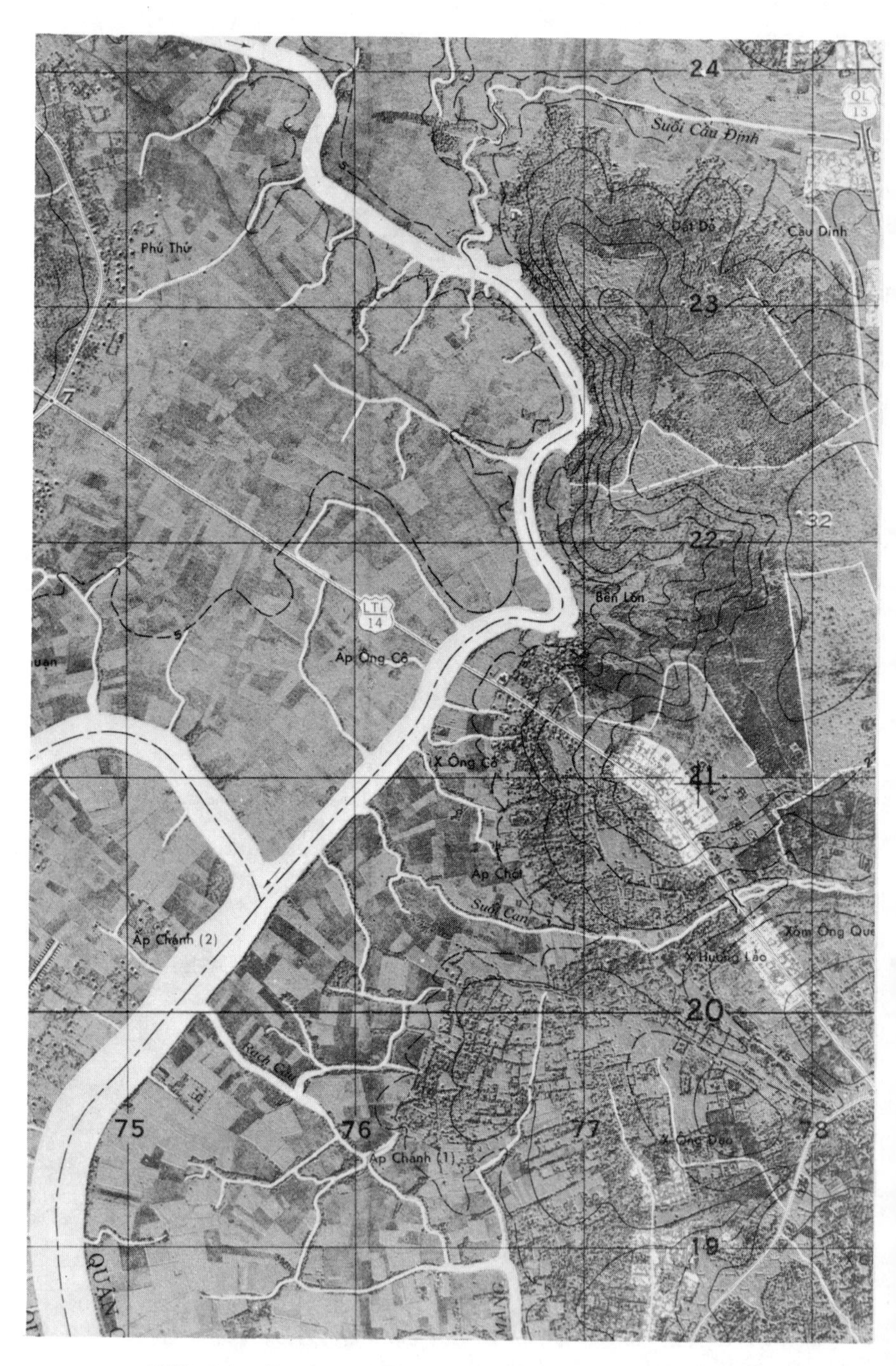

FIG. 8.4. *Portion of Pictomap (original scale 1:25,000) of part of Viet Nam. Prepared by the Army Map Service, later TOPOCOM, and now the Topographic Center, Defense Mapping Agency.*

annotated mosaics of air photos are used to show the character of areas; in some series the air photo coverage is printed on the reverse side of a topographic sheet at approximately the same scale as the map. In addition, rectified, controlled photo mosaics are sometimes used as maps with annotations overprinted in color, as in the Pictomap of the Topographic Center, D.M.A. (Fig. 8.4) which is similar to orthophotomaps of the U.S.G.S. Such presentations are particularly valuable when the nature of the surface cover, e.g., vegetation, swamps, etc., is important to the user.

To illustrate the character of topographic maps we have reproduced the Orbisonia, Pennsylvania sheet from the United States Geological Survey (U.S.G.S.), 1:62,500 series. Figure 8.5 shows this sheet in contoured form, and Figure 8.6 is a contoured and shaded relief version of the same map for comparison. It is unnecessary to go into detail concerning symbolization on topographic maps but, in general, water features are rendered in blue, cultural features in black and red, and relief in brown. Sometimes broad classes of vegetation are added in green. A 1:62,500 U.S.G.S. quadrangle covers 15′ (fifteen minutes) of latitude by 15′ (60′ = 1°) of longitude characteristically and has a contour interval of 20 feet. Other sheet margins and other contour intervals are used for particular situations. For example, in areas of very low relief a 5 foot contour interval may be used and some special sheets, such as those of the National Parks, may cover more than 15′ × 15′ to accommodate the whole of such an area on one map.

In the printing of colored maps, especially topographic maps, photolithography is now the usual method of reproduction. The cartographer prepares plates, one for each color, from which negatives can be made. Since World War II a method known as plastic scribing has been increasingly used for this purpose. In this process, the cartographer uses a dimensionally stable plastic sheet which has a photographically opaque coating applied to it. He scribes, i.e., scratches through the coating, the desired information and in this way prepares what amounts to a negative, so that the photographic negative stage can be by-passed; obviously positive images can also be produced from the negatives. From the scribed sheets (one for each color), the metal printing plates, which have replaced stone formerly used, can be prepared directly. These metal plates are then introduced into a rotary press which allows one color to be printed after the other with great speed. Plastic scribing has returned control of the production of maps more particularly from the photographer to the cartographer.

In addition to the conventional 1:62,500 contoured topographic quadrangles, the U.S.G.S. prepares maps of other types and scales. Thus, as illustrated (Fig. 8.6), shaded relief has been applied to some sheets to give a remarkably plastic effect. Contouring supplemented by shaded

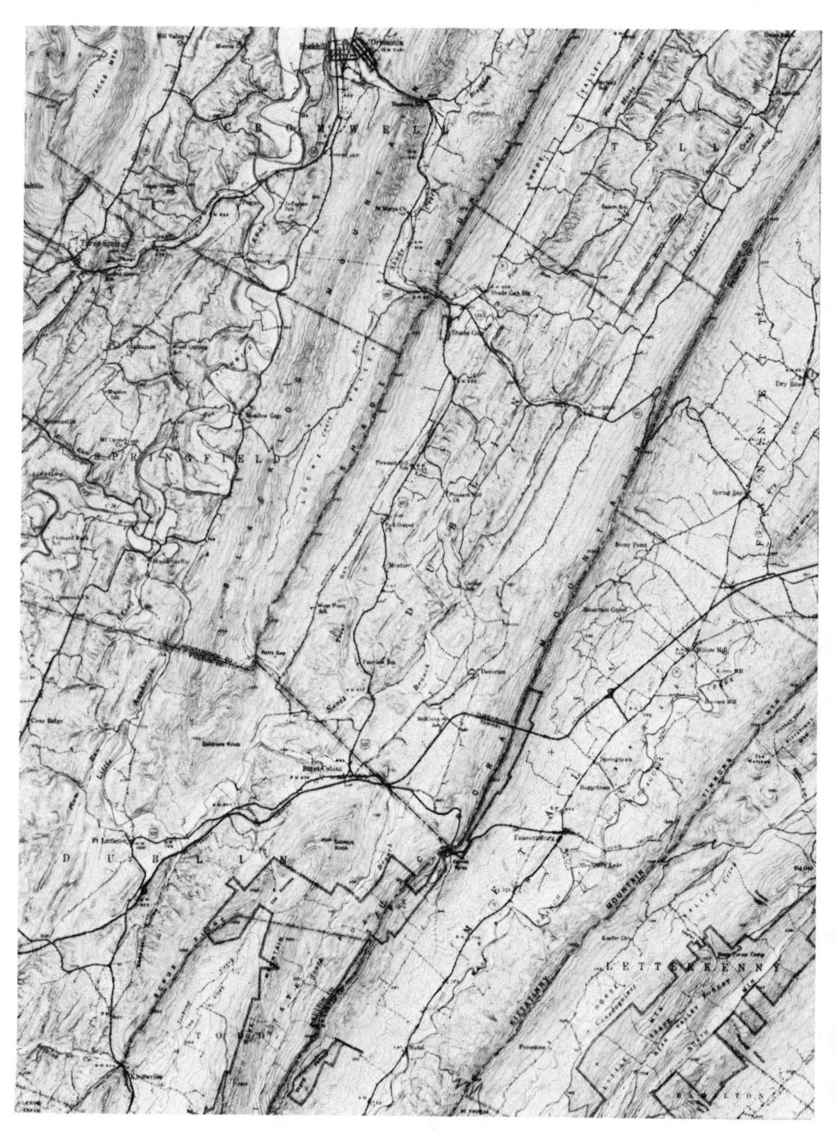

FIG. 8.5. *Contoured topographic map of Orbisonia, Pennsylvania; scale of original 1: 62,500 by the United States Geological Survey.*

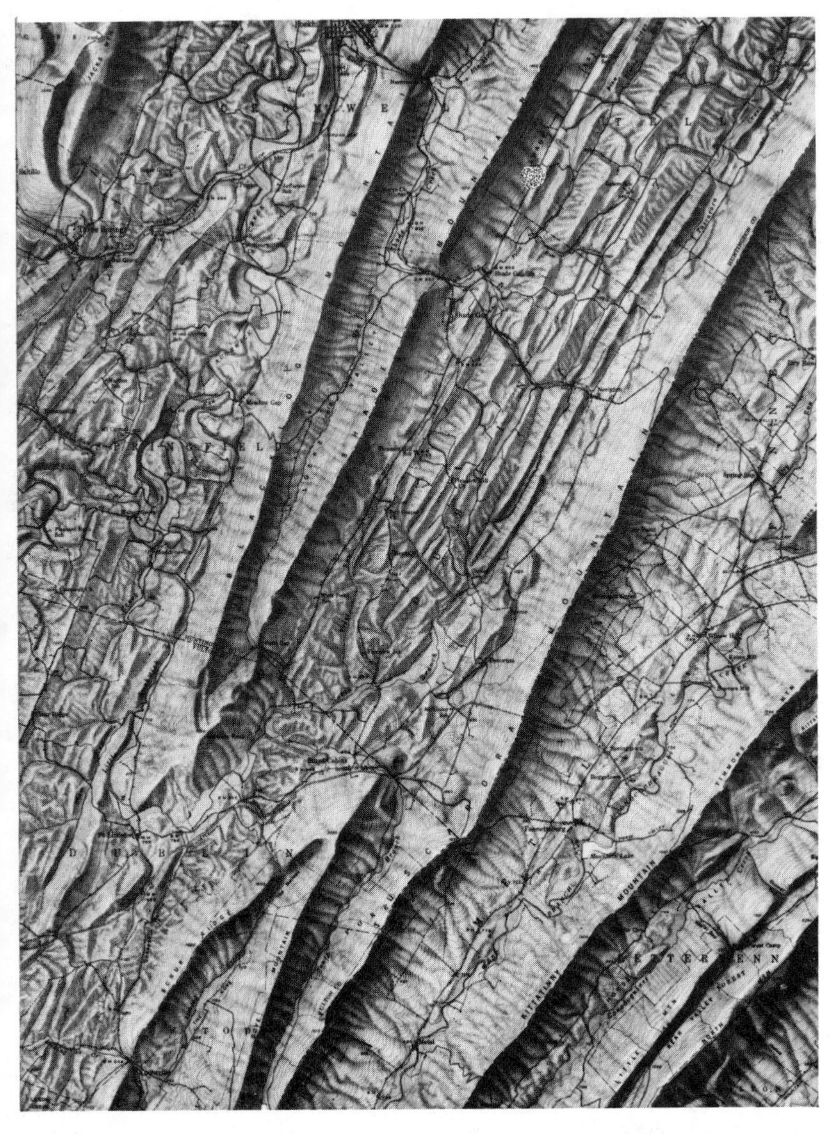

FIG. 8.6. *Shaded relief edition of contour map of the area shown in Fig. 8.5, by the United States Geological Survey (both reduced).*

relief (now often rendered with an air brush) approaches the ideal in cartographic representation of terrain, combining, as it does, a quantitative method with one that is easy to comprehend visually. Obviously, however small the contour interval, important details that may give real character to an area and which are observable on the air photos can be lost between the contours. One may imagine, for example, what would be lost in the representation of a man with a 5 foot contour interval. On the other hand, small but significant features can be rendered in the continuous tone of the plastic shading. Interestingly, the landscape shown in Figure 8.6 is shaded as though illuminated from the northwest, a direction from which the area actually never receives its light. This convention has a practical basis since illumination from the bottom of the map produces a pseudoscopic (inverted image) effect. The U.S.G.S. also makes topographic maps of more detailed scale than 1:62,500, particularly the 1:24,000 (7½′ × 7½′) series for urban areas and other places that require coverage for purposes such as engineering, urban planning, etc. In fact, for areas covered by this series, the 1:24,000 maps have now essentially replaced the 1:62,500 sheets. For regional planning, the U.S.G.S. publishes the National Topographic Maps of the United States at a scale of 1:250,000. These and other maps published by United States government agencies are in the Public Domain and therefore are not copyrighted, as is common in most foreign countries. To illustrate the magnitude of the task of mapping a country as large as the United States, after nearly a century of splendid work, the U.S.G.S. has mapped about a quarter of the area with sheets of standard quality. The rest of the coverage is by reconnaissance, out-of-date or other less-than-first-class quality maps. One of the problems is that the U.S.G.S. is not the sole authority for domestic topographic mapping in the United States in the sense that the Ordnance Survey is in Britain; the duties of topographic mapping in the United States are shared between a number of government agencies. Another problem is that topographic mapping, except in special circumstances, is initiated by the States and financed on a shared cost basis. In many States mapping has a low order of priority.

The United States Geological Survey was founded in 1879 as the result of the remarkable mapping activities associated with the detailed exploration and opening up of the arid American West. Its purpose was to make "a systematic study of the geology and natural resources of the United States and for classifying the public lands."[6] The foundation of

[6] United States Geological Survey, "Topographic Maps—Descriptive Folder" (Washington, D.C.: U.S.G.S., regularly revised). The U.S.G.S., like other great surveys of the world, publishes status maps of topographic coverage, air photo coverage, geodetic control, etc.

the U.S.G.S. thus differed from the older topographic surveys in Europe which were more strictly geographical, topographical, or military in outlook. Having been brought into being after the science of geology was well developed and in accordance with the terms of its establishment and title, the U.S.G.S. in its early years paid particular attention to the physical landscape. For example, the relief rendering on early U.S.G.S. maps is among the best in the world for the time; less importance was attached to cultural features in comparison with the maps of a number of other surveys.[7] However, the increasing importance of the works of man upon the land has caused a change to the point where U.S.G.S. maps, particularly those of the 1:24,000 series, are now much richer in cultural detail. The U.S.G.S. maintains a Map Information Office in Washington, D.C. to provide data on maps, air photos and geodetic control in the United States, its territories and possessions.

Only a fraction of the land surface of the earth has been covered by up-to-date, topographic maps suitable for engineering purposes.[8] Generally European countries are well mapped topographically. France and Britain were surveyed uniformly but, because of political fragmentation until the middle of the last century, Germany and Italy did not have centralized control of surveys when they were initiated and this later caused problems in the mapping of those countries. Former colonial areas are often surprisingly well covered by topographic maps; for example, India is much better mapped today than China. The Directorate of Overseas (formerly Colonial) Surveys (D.O.S.) was established in Britain to assist certain developing countries in their topographic mapping endeavors. Most of the work is now done near London by air survey methods. The Institut Géographique National of France serves a similar function to the D.O.S. It can be said, with truth, that we do not really know an area (resources, morphology, size, etc.) until it has been mapped in detail.[9] Furthermore, many human activities, including the compilation of other kinds of maps, cannot be accomplished without good topographic coverage.

[7] Everett C. Olson and Agnes Whitmarsh, *Foreign Maps* (New York: Harper, 1944) illustrates by small samples, some in color, major sheet maps of the world. See also C. B. Muriel Lock, *Modern Maps and Atlases: An Outline Guide to Twentieth Century Production* (Hamden, Conn.: Archon Books, 1969). This work, unlike the Olson and Whitmarsh volume, lacks illustrations except for a frontispiece.

[8] H. Arnold Karo, *World Mapping, 1954–55* (Washington, D.C.: Industrial College of the Armed Forces, 1955). Although now somewhat out of date, this book has an excellent series of world "appraisal" maps of triangulation, coverage of topographic maps, nautical charts, aeronautical charts, etc.

[9] Map reading, measurement, analysis, intelligence, etc., have been the subject of a considerable literature, most of which emphasizes the topographic map. Good representative examples are: Department of the Army, *Map Reading* (Washington, D.C.: Department of the Army Field Manual 21–26, March 1965); T. W.

We have mentioned geological and land use maps as two classes which are fundamentally dependent on the topographic base. Detailed geological mapping is generally performed on a piecemeal basis according to need. One of the most remarkable mapping enterprises in recent times, covering a country, is the Land Utilisation Survey of Great Britain conducted under the direction of Professor (later Sir) L. Dudley Stamp. The principal aims of this survey, which came into being in 1930, were "to make a record of the existing use of every acre in England, Wales and Scotland," and to "secure the support of a well informed public opinion for the work of planning the land for the future."[10] A number of the most important individuals involved in economic and environmental planning in Britain in the 1930's were connected with the project, but the field work was accomplished largely by secondary school students under the supervision of their geography teachers. The compilation scale was 6 inches to 1 mile; the sheets were reduced to 1 inch to 1 mile (1:63,360) and published on the Ordnance Survey base maps of this scale (Fig. 8.7). Land uses are indicated by colors as follows: forest and woodland (dark green); meadow and permanent grass (light green); arable land (brown); heathland, etc. (yellow); gardens and orchards (purple); buildings, yards, roads, etc. (red). This color scheme has now become more or less conventional for land use maps, just as the topographic map color scheme indicated earlier is conventional for maps of that type. By 1940 the Land Utilisation Survey of Britain was virtually complete and proved of enormous value as Britain systematically expanded its agricultural production in World War II. The recent British experience in land use mapping has had several important results: (*1*) it has provided a base of comparison with past and future land uses; (*2*) it has given rise to even more detailed (1:25,000) land use mapping activities in Britain; and (*3*) it has led to the formation of a commission of the International Geographical Union appointed to apply the principles to other countries, especially developing nations. Land use mapping

---

Birch, *Maps, Topographical and Statistical,* 2nd ed. (Oxford: The Clarendon Press, 1964). An interesting departure from such traditional map reading manuals is Armin K. Lobeck, *Things Maps Don't Tell Us* (New York: Macmillan, 1956). David Greenhood, *Mapping* (Chicago: University of Chicago Press, 1964) deals with the technique of map-making from a topical point of view. Norman J. W. Thrower and Ronald U. Cooke, "Scales for Determining Slope from Topographic Maps," *The Professional Geographer,* vol. 20, no. 3 (1968), 181–86. One hundred quadrangles were specially selected from the U.S.G.S. topographic coverage of the United States to illustrate physiographic types. See also *Rural Settlement Patterns in the United States as Illustrated on 100 Topographic Quadrangle Maps,* National Academy of Sciences, National Research Council, Publ. 380, 1956.

[10] L. Dudley Stamp and E. C. Willatts, *The Land Utilization Survey of Britain,* 2nd ed. (London: London School of Economics, 1935). A verbal description of the survey was published in ninety-two parts between 1937 and 1941.

THE LAND UTILISATION SURVEY OF BRITAIN

Specimen of a " One-Inch " map

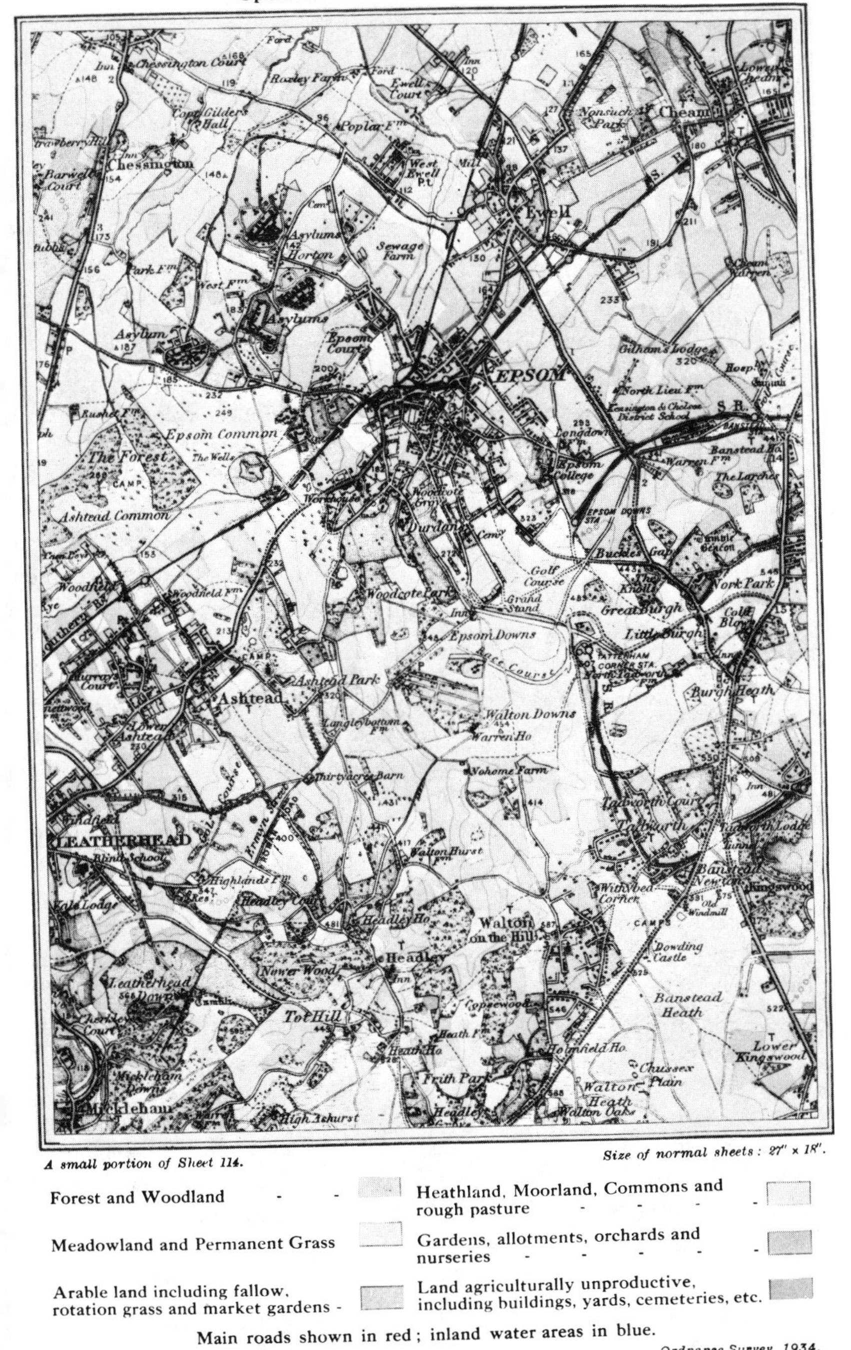

FIG. 8.7. *Specimen sheet with legend of Land Utilisation Survey of Britain; scale of original 1: 63,360.*

of chorographic scale has been accomplished exclusively with the aid of Gemini and Apollo satellite imagery; because of the broad view it affords, some relationships are better appreciated on space photography than by any other means.[11]

Perhaps the most familiar official map to many people is the Daily Weather Map which is reprinted in simplified form in newspapers. The United States Daily Weather Map originates with the Environmental Science Services Administration (ESSA)/Environmental Data Service (formerly the U.S. Weather Bureau), Department of Commerce. ESSA maintains over 12,000 weather stations of which 200 are classed as first order stations. These last, located in major cities and at airports, are staffed with professional forecasters who furnish meteorological data in the form of daily reports. Every 24 hours the National Weather Analysis Center in Washington, D. C. receives 22,000 hourly surface reports, 8,000 international 6-hourly surface reports, as well as a smaller number of reports from ships, balloons, etc. Transmissions are made in abbreviated numerical code by radio, telegraph, or teletype. These data are then plotted on maps. The Daily Weather Map shown in Figure 8.8 was compiled from observations made at 7:00 a.m. Eastern Standard Time, Thursday, January 1, 1970.

In discussing the symbolization on the Daily Weather Map, it is well to recall that this device, of value to almost everyone in the country, came into being after a long period of gestation. We have already discussed the wind chart developed by Halley and isolines applied to climatological phenomena by Humboldt and others. Among the earliest climatological maps on which a variety of phenomena appear together on one chart are those of the Americans, Elias Loomis in the 1840's and Commodore Maury in the next decade. These maps show a composite of weather over a considerable period of time, differing in this respect from the Daily Weather Map which is a "snapshot" of weather conditions. Nevertheless, the symbols on contemporary weather maps developed from earlier climatological map symbols, e.g., arrows used for wind direction, isolines for barometric pressure (isobars), etc., according to an international code. The arrow symbol has been refined to show not only wind direction, but wind velocity according to a scale developed by, and named after, the English hydrographer Sir Francis Beaufort (1774–1857). This is accomplished by "feathers" on the arrows, the number and length

[11] Norman J. W. Thrower assisted by Leslie W. Senger and Robert H. Mullens, cartography by Carolyn Crawford and Keith Walton, "Land Use in the Southwestern United States from Gemini and Apollo Imagery," Map suppl. no. 12, *Annals of the Association of American Geographers,* vol. 60, no. 1 (March 1970).

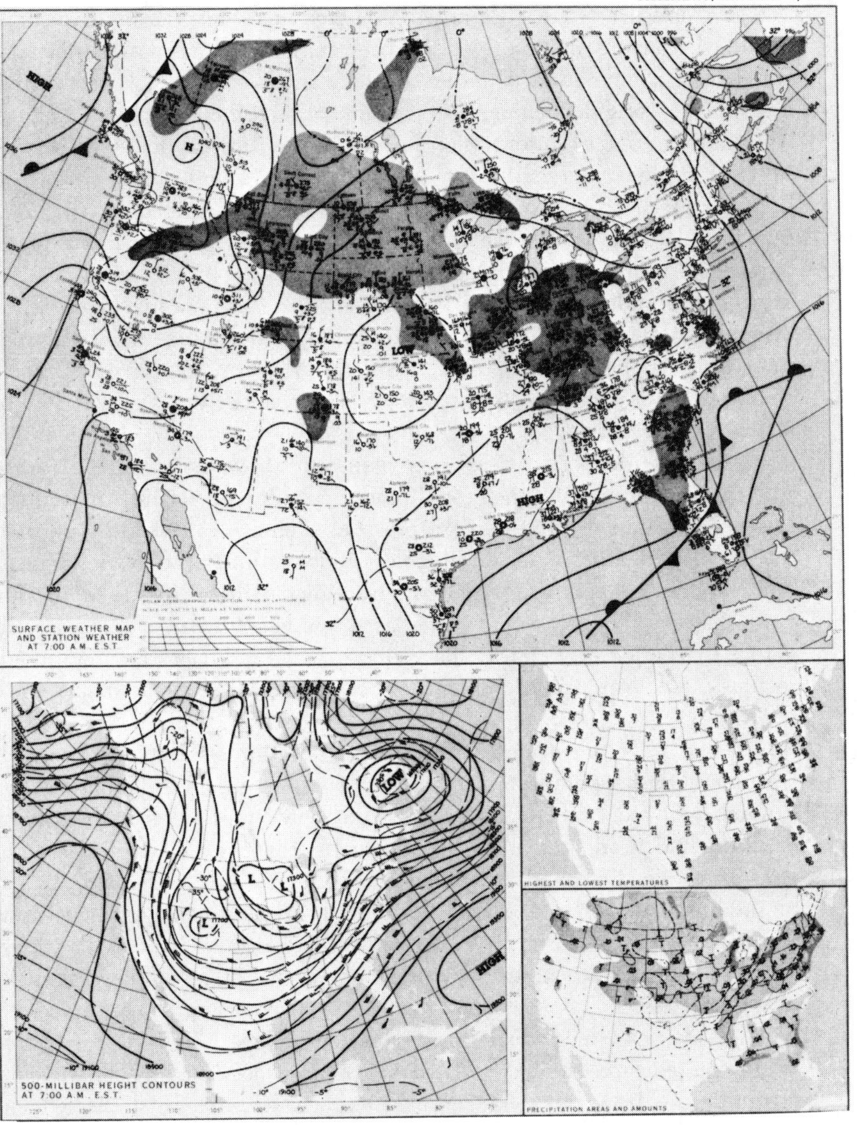

FIG. 8.8. *Summary sheet of Daily Weather Map information for Thursday, January 1, 1970.*

of which indicate the speed of the wind in knots. Isobars (with pressure reduced to sea level) are drawn at intervals of four millibars, and high and low pressure centers are noted. Another line symbol is used to show fronts whether warm, cold, occluded, or stationary. Air masses are indicated by a combination of letters which suggest moisture characteristics and area of origin. Temperature, cloud, and visibility data as well as wind and barometric information appear in the form of point symbols at selected stations scattered over the map. Areal symbols are used for precipitation, rain and snow being differentiated. Theoretically some 200 symbols might appear on the Daily Weather Map which is a masterpiece of condensation. The completed map is photo-electrically produced in Washington and relayed to stations which have facsimile machines capable of reproducing the original. In addition to the four daily (6-hourly intervals) synoptic charts, the machines process prognostic maps, constant level and constant pressure charts of the upper air, and wind-aloft charts.[12]

A sequential series of weather maps can show the passage of pressure cells and fronts and the growth and dissipation of storms. An attempt to make this information more graphic has been accomplished by putting the 6-hourly weather maps on film to produce a time-lapse movie. This is one step toward animated cartography which will be discussed in the next chapter. The use of satellites as a source of weather data began before the period of manned space flight. Satellite images are now used to illustrate storms and cloud phenomena, with and without map annotations.

Many government departments other than those specifically charged with mapping have important cartographic sections. In the United States a number of federal agencies produce not only maps to serve their own departmental needs but also thematic maps which are of interest to the general public. Outstanding among these are small-scale thematic maps of the entire United States by the U.S. Department of Agriculture and the Bureau of the Census. The Agriculture Department has a long tradition of support to thematic mapping and several of its employees, such as the late Oliver E. Baker and Francis J. Marschner, must be numbered among America's most distinguished cartographers. The series of maps of the United States on the scale of 1:5,000,000 published by the Bureau of the Census from the 1960 Census cover a wide variety of topics which

[12] *Daily Weather Map* (Environmental Science Services Administration/Environmental Data Service, U. S. Department of Commerce), has an explanation of symbolization on the reverse of the sheets. *Daily Weather Maps, Weekly Series,* with special explanatory sheet, are a continuation of the Weather Bureau publication, *Daily Weather Map.* A useful general classification of map symbols of all kinds was made by John K. Wright, who categorized point, line, surface, and volume symbols and further subdivided these. See footnote 5, Chap. 9.

arise from the census returns.[13] Such maps, and those from other government departments in the United States and elsewhere find important use in the classroom as mounted maps or as reference items in sheet map collections.

In the United States most cadastral mapping now, as at earlier periods, is accomplished at the local level—county, civil township, city, etc., and much of the record is in manuscript form. The lack of comparability in the information is particularly critical in today's cities with their pressing problems. A plea has already been made for the co-ordination of city surveying and mapping activities in the United States. It is pointed out that in a given city the interdepartmental and private demands for map information often involve a serious wastage of time through lack of a centralized cartographic clearing house.[14] The situation between different cities, in this respect, is nothing short of alarming, especially in those cases where one urbanized area abuts another. Political, legal, and financial aspects of municipal surveying and mapping demand more co-ordination of these activities. The suggestion has been made that, at least, a centralized city survey and mapping agency should be founded to establish standards, to educate both surveyors and the public, and to eliminate wasteful duplication of effort. The problems in Great Britain are much simpler but, in any case, in that country large scale plans of 1:2,500 (approximately 25 inches to 1 mile) are available for the entire Kingdom, except some areas of mountain and moorland. In addition Britain has new plans at twice this scale (1:1,250) of most built-up areas, a work to be completed by 1980. Such maps include all municipal boundaries, roads, alleys, public buildings (named), houses (numbered) and even smaller structures. To print and keep up-to-date such maps is a formidable task, but in a planned society like Britain's, the effort is regarded as well worthwhile because of the value of the maps to administrators and others.[15]

The *Atlas of Finland* is usually considered to be the first true national atlas with a continuous existence as indicated earlier. However,

[13] The United States Department of Agriculture published general U. S. thematic maps on Land Use, and Predominant Economies, etc., while the Bureau of the Census, Department of Commerce, produced a series of maps arising from the 1960 Census covering such topics as Population, Housing, Income, etc. Toponymy (geographical nomenclature) in the United States is officially the concern of the Board on Geographic Names of the Department of the Interior.

[14] Joseph M. Dearborn, "The Co-ordination and Administration of City Surveying and Mapping Activities," a paper presented at the Sixth Annual Meeting, American Congress on Surveying and Mapping, Washington, D.C., June 28–29, 1946.

[15] Richard A. Gardiner, "How Britain's National Maps are Produced," *The Penrose Annual,* vol. 57 (1964), 211–18.

a claim is made for the Royal Scottish Geographical Society's *Atlas of Scotland* published by the private map publishing house of J. G. Bartholomew in 1895 as being the first national atlas.[16] In a case like this, it is sometimes difficult to determine whether an atlas of a country, group of states, state, or province is official, quasi-official, or private. An atlas might be sponsored by a government department, compiled at a university, and printed by a private company. At all events increasing numbers of atlases dealing with discrete political units have appeared in recent years. Some of these focus on a particular set of data within the selected area, for example the *Atlas of American Agriculture* by Oliver E. Baker (1936). More often these national or state atlases deal with a very wide range of information as, for example, the recently (1970) completed *National Atlas of the United States of America* prepared under the direction of Arch C. Gerlach, Chief Geographer, United States Geological Survey, Department of the Interior. National atlases have been published by technologically advanced countries, e.g., Canada and France, and also by developing ones such as Kenya and Uganda. The maps in the *Atlas of Finland* covering, as they do, various aspects of geology, climate, vegetation, population, agriculture, forestry, industry, transportation, trade, finance, education, health, elections, as well as political/historical data and geographic regions may be taken as representative of the contents of an excellent national atlas. Naturally the subjects selected are those which are important to the particular country represented and no satisfactory format for all national atlases should or could be devised. There is a great variety in the quality of the maps found in national atlases, depending on the reliability of the census data, the ingenuity of the cartographers, and the skill of the printer.[17]

Many of the above remarks are true of atlases of regional, state, and provincial areas. Again, commercial concerns, universities, and gov-

[16] Gerald R. Crone, *Maps and Their Makers: An Introduction to the History of Cartography* (London: Hutchinson, 1966), p. 169.

[17] Countries with national atlases include those on the list below in which the name of the country is followed by the date of the initial appearance of the work. Recent editions are indicated in parentheses, e.g. (3rd 1957); those appearing sheet by sheet thus, 1950–; and economic atlases in italics: Finland 1899 (4th 1960); Canada 1906 (3rd 1957); Egypt 1928; Gold Coast (Ghana) 1932 (2nd 1949); France 1934 (2nd 1953); Czechoslovakia 1935 (2nd 1966); Ukraine 1937; Netherlands East Indies (Indonesia) 1938; British Honduras (Belize) 1939; U.S.S.R. (Atlas Mira) 1939; *Italy* 1940; *Norway* 1940; Portugal 1941 (2nd 1958); Tanganyika (Tanzania) 1942 (3rd 1956); Dominican Republic 1944; Belgium Congo (Republic of the Congo) 1948; Portuguese Colonies 1948 (Mozambique 1960); Denmark 1940–; Mexico 1949; Belgium 1950–; *West Germany* 1950–; Australia 1952; Yugoslavia 1952; Costa Rica 1953; Poland 1953–; Sierra Leone 1953; Sweden 1953–; Morocco 1954–; Argentina 1954; *Japan* 1954; *El Salvador* 1955; Spanish Africa 1955; U.S.A. 1956–; Israel 1956

ernment departments are involved in the production of such works. Thus the *Atlas of the State of South Carolina* (1825), the first such project in the United States, was authorized by the State Legislature, compiled by an individual, and printed commercially. Increasingly, universities, chambers of commerce, and state and provincial governments have concerned themselves with such ventures. At the present time about two-thirds of the states, provinces, and territories of the United States and Canada have a specific thematic state atlas. Outstanding among these works are the *British Columbia Atlas of Resources* and the *Economic Atlas of Ontario*.[18] These atlases present a wealth of information about the resources of their respective provinces. They should serve as models for states and provinces lacking such coverage or with inadequate atlases.

We should not conclude this chapter on contemporary official cartography before discussing the progress of extraterrestrial mapping. Pioneer work in this field by Galileo, and the celestial charting activities of Halley and Hevelius have been mentioned. These scientists were followed by a succession of astronomers who interested themselves in lunar mapping, an activity which, in its development, can be considered a microcosm of geo-cartography, albeit with some important differences. Most of the early lunar maps were drawn on the orthographic projection in which a sphere appears as viewed from an infinite distance. Such is the case of the moon map of Franciscus Fontana (1600–1650) who was also the first man to observe the markings on Mars and to sketch this planet. In this period the usual method of representing the relief of the moon was by line shading, although Hevelius had used "mole hill" forms

---

(2nd 1962); India 1957; Byelorussia 1958; Brazil 1959 (1966); China (Republic) 1959–; *Colombia* 1959; Kenya 1959; New Zealand 1960; South Africa 1960; Austria 1961–; Uganda 1962; *Spain* 1963; Panama 1965; Chile 1966; *Greece* 1966; Hungary 1967. This list does not include atlases which are essentially national atlases but where the sponsoring or issuing authority is not a government department *viz. Turkey* 1961; Britain 1963; Rhodesia 1965; Senegal 1965. Data contained in *The Atlas of Britain,* (Oxford: Clarendon Press, 1963), p. VII has been up-dated for this footnote. In that atlas, as in many others, whether specifically designated as economic or not, economic maps often form an important part of the coverage. An example of a supra-national atlas in terms of the area covered is the Soviet, *Atlas of Antarctica,* vol. 1 (Moscow, 1966). For state and provincial atlases, official and private, of North America, see Richard W. Stephenson and Mary Galneder, "Anglo-American State and Provincial Thematic Atlases: A Survey and Bibliography," *The Canadian Cartographer,* vol. 6, no. 1 (1969), 15–45.

18 J. D. Chapman and D. B. Turner, eds., and A. L. Farley and R. I. Ruggles, cartographic eds., *British Columbia Atlas of Resources* (Vancouver: British Columbia Natural Resources Conference, 1956); and William G. Dean, ed., and G. J. Matthews, cartographer, *Economic Atlas of Ontario* (Toronto: University of Toronto Press and the Government of Ontario, 1968).

on one of his maps.[19] The common system of nomenclature for lunar features, i.e., naming after great men, was devised by Jean Riccioli in the 1650's, and to a large extent displaced Hevelius' method of using the names of prominent earth features for the purpose, although some of this terminology is still in use. Like Hevelius, Riccioli showed the moon's librations (oscillations which enable us to see more than half of the lunar surface from the earth) by intersecting circles. J. D. Cassini made lunar maps in the 1680's but these represented no real advance over the work of Hevelius. In the eighteenth century, Johann Mayer used a micrometer to measure lunar elevations and locations and devised a net of equatorial coordinates for the lunar surface. J. H. Lambert (see p. 82 for his work on projections) employed a conformal projection—the Stereographic, for lunar mapping; this allowed the features, including peripheral ones, to be shown with a minimum of distortion. Of course, a distant view of the subject with its obvious advantages has always been available to lunar map-makers, but only recently have cartographers had a similar advantage for terrestrial mapping through aerial and later satellite photography. Two of the greatest selenographers of the nineteenth century were Johann H. von Mädler and Wilhelm Beer who collaborated on a lunar chart in which they used triangulation. They employed hachures to represent the lunar relief while, later in the century, L. Trouvelot used a double or triple circle (form line) method. This was a step toward a contour map of the moon which was made possible through lunar photography which began in the 1850's. Because of the moon's librations, stereoscopic pictures of this body can be taken. However, it was not until 1920 that H. N. Kempthorne demonstrated that the principle would permit satisfactory contour mapping of lunar features, a work carried forward by Phillip Fauth in the 1930's. Opposition to the use of the contour for selenographical purposes had to be overcome, but this method combined with shaded relief, or with altitude tints between contours, has now become the usual means of representing the surface of the moon.

[19] Judith A. Zink (Tyner), "Lunar Cartography: 1610–1962" unpublished M. A. thesis (University of California, Los Angeles, 1963), and "Early Lunar Cartography," *Surveying and Mapping,* vol. 29, no. 4 (1969), 583–96; and W. F. Ryan, "John Russell, R. A., and Early Lunar Mapping," *The Smithsonian Journal of History,* vol. 1 (1966), 27–48. Russell (1745–1806), a prominent artist produced excellent shaded relief renderings of the moon (for which he gathered data by serious astronomical observation) in pastel and other media. In his time there was no adequate means of reproducing these continuous tonal drawings directly; indeed, before the development of modern reproduction techniques, the artist, like the cartographer, had to rely on the interpretation of an engraver for multiple copies of his work unless, of course, he did his own engraving as was true in some cases.

In very recent years, since the first manned space flights and the return of pictures of the unobserved side of the moon from an unmanned Russian satellite, there has been a great flurry of interest in lunar mapping. While previously individual scientists labored for a lifetime to produce a representation, now teams of cartographers from large official map-making agencies are engaged in the work. In the United States, for example, since the late 1950's the U.S.G.S. and various mapping agencies of the United States military establishment have been active in lunar mapping. Without attempting to list all of the lunar maps produced by these agencies, it can be said that they range from single sheets of small scale, to multi-sheet series at 1:1,000,000 [such as the Lunar Aeronautical Charts (Aerospace Center)] and larger scales. A number of projections other than the Orthographic and Stereographic are now used for lunar mapping, including the Lambert Conformal and Mercator projections. The subjects covered include relief, physiography, tectonics, surface materials, and "geology." Purists have taken exception to this last term being applied to the lunar lithology and have preferred the expression, "selenology." More general terms such as "cosmology" and "cosmography" may have to be revived; happily "cartography" and "map" are general terms equally applicable to terrestrial or extraterrestrial phenomena. As

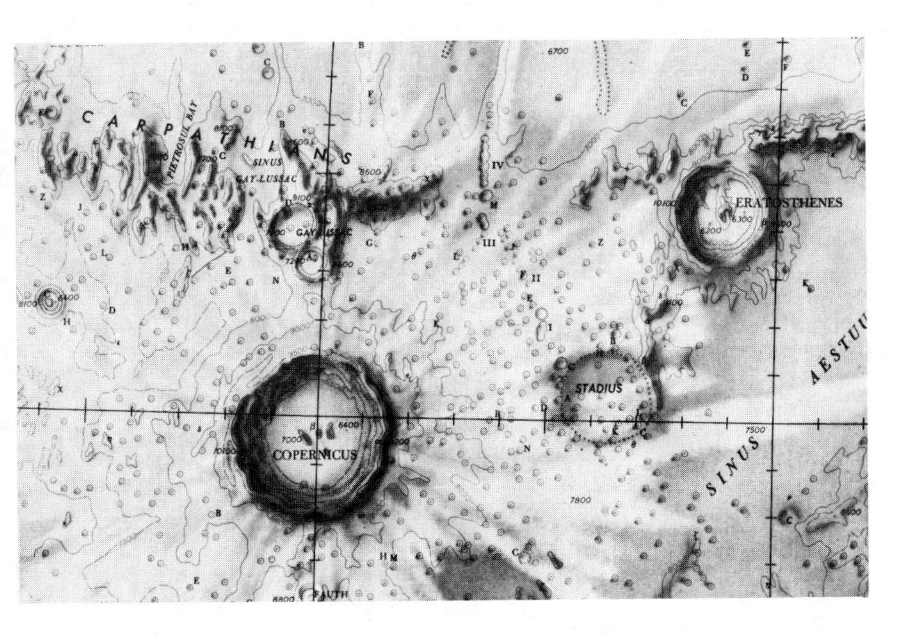

FIG. 8.9. *Small section of a lunar chart (reduced), scale of original 1:2,500,000 from the Topographic Center, D.M.A.*

indicated, a variety of techniques have been used to express the surface features of the moon—shaded relief, contours, hypsometric tints, etc. As in terrestrial mapping, the combining of these methods is facilitated through color printing. A problem in expressing the lunar surface by contours has been the lack of a "natural" datum such as mean sea level on the earth. Frequently the center of the crater, Mösting A, is utilized for the vertical datum for contouring and thus a point, rather than a surface, is used for this purpose. Celestial bodies other than the moon also command the attention of modern cartographers. Recently a beautiful map of Mars at the scale of 1:10,000,000 on the Orthographic projection based on the notes of a number of astronomers has been prepared by the Topographic Center, Defense Mapping Agency.

Figure 8.9 is used to illustrate contemporary extraterrestrial mapping. It shows a small part of the surface of the moon centering on the crater Copernicus from the D.M.A. Lunar Map on the scale of 1:2,500,000. A combination of contouring and shading is employed for the relief representation. Both regular and depression contours are evident on the larger craters, while small craters are marked by uniform circles. Prominent ridges of less than the contour interval (1000 meters with supplementary contours at 500 meters) are marked by a series of dots as at Stadius. Color is used to distinguish the rays (light), craters (yellow to brown), and maria (green to blue). The map is drawn on a Modified Stereographic projection, related to a selenodetic sphere, relief data being compiled by stereophotogrammetric methods.

In the preceding consideration of certain highlights of official cartography in the present century we noticed that in some instances it is difficult to make a clear-cut distinction between official and private mapping activities. By the same token, some private cartography is subsidized to such an extent by government funds that it is questionable whether it can be properly designated as unofficial. Governments, particularly national governments, are major patrons of scientific activity today as individuals, especially rulers, were at an earlier period. In one way or another, government grants undergird much cartographic activity in universities and professional societies. Such institutions often have cartographic staffs which, particularly in time of national emergency, have been engaged extensively in government contract work. Nevertheless, as a matter of convenience the chapters in this book on modern cartography have been divided between predominantly official mapping on the one hand, and predominantly private on the other. Ideally in private cartography, the researcher initiates the project, receives individual credit for it, and retains responsibility for the work even though it may be supported by public funds. Specific credit for maps is sometimes given for government cartographic work but the project is the responsibility of the department concerned, as a general rule.

In addition to the maps on the pages of their periodicals, various private geographical societies provide readers with occasional folded map supplements of larger than page-size formats. This is a well-established tradition in Europe, and in the United States, *The Geographical Review, The Annals of the Association of American Geographers,* and the *National Geographic Magazine,* for example, all publish map supplements.

Modern Cartography: Private and Institutional Maps

9

Such maps perform a valuable function since they show distributions which cannot adequately be represented on the pages of a journal and complement, rather than duplicate, map-making activities of other organizations. Taken altogether the cartographic output of such societies is impressive, varied, and often experimental in nature. Sheet maps illustrating demographic, ethnographic, linguistic, geological, biogeographical, medical, historical, and a wide range of other distributions have been made under the auspices of professional societies.

To illustrate this type of cartography, a small sample of a population map of California by the author appears as Figure 9.1. This map, published in the map supplement series of the *Annals of the Association of American Geographers,* was produced as part of the work of the Commission on a World Population Map (now the Commission on the Cartography and Geography of World Population) of the International Geographical Union.[1] The map utilizes areal symbols and open circles, proportional to population, for cities; larger cities are keyed by letters to a list of places and their population. Dots with a unit value of 50 persons are used for rural population. The value, size, and location of dots, all present special problems to the cartographer. The original of this map utilizes red tone for urban areas, red dots for rural population, and black circles and lettering for cities with a yellow shaded relief background to help explain demographic concentration and dispersal. It is an attempt to solve a central problem of demographic mapping, namely to represent on one map the great difference in population density between urban and rural areas.

Simulated three-dimensional symbols—spheres (developed by the well-known Swedish geographer, Sten de Geer), cubes, etc., have also been used extensively for population and economic mapping. However, a recent psychometric study suggests that these simulated volumetric symbols "merely reflect their perceived area" and "are not efficient in creating the desired impression of volume."[2] A number of other cartographic symbols—graded patterns, circles, dots, etc., have been subjected to rigorous analysis in recent years.[3] All too few adequate maps exist

[1] W. William-Olsson, "Report of the I.G.U. Commission on a World Population Map," *Geografiska Annaler,* vol. 45, no. 4 (1963), 243–91, consists of a series of articles on different aspects of population mapping.

[2] Gosta Ekman, Ralf Lindman, and W. William-Olsson, "A Psychophysical Study of Cartographic Symbols," *ibid.*, 262–71.

[3] Published and unpublished studies on individual symbols used in thematic cartography appearing in recent years include: Robert L. Williams, "Equal-Appearing Intervals for Printed Screens," *Annals of the Association of American Geographers,* vol. 48 (1958), 132–39; George F. Jenks and Duane S. Knos, "The Use of Shaded Patterns in Graded Series," *Annals of the Association of American Geographers,* vol. 51 (1961), 316–34; and J. Ross MacKay, "Dotting the Dot Map," *Surveying and Mapping,* vol. 9 (1949), 3–10; Richard E. Dahlberg,

FIG. 9.1. *Small section of map "California Population, 1960," focusing on the San Francisco area. Scale of original 1: 1,000,000.*

of population, a phenomenon as important to cultural geography and social science generally as relief, climate, etc., are to physical science. On the whole, cartographic methods to serve the physical sciences are better developed than techniques used for social sciences, but the latter are now receiving a good deal of attention.[4] This order of development follows the general progress of science in which, by and large, the physical sciences reached a high level of sophistication earlier than those studies which deal systematically with life and man.

Usually the sheet maps issued by private societies are individual efforts which do not relate specifically to other maps; they may represent the culmination of many months or even years of research. However, sometimes series of maps are sponsored privately such as the previously discussed "Map of Hispanic America on the Millionth Scale" issued by the American Geographical Society. The same society published another outstanding series of maps between 1950 and 1955 showing the World Distribution of Cholera, Dengue and Yellow Fever, Leprosy, Malaria, Plague, Polio, etc., the work of Dr. Jacques May.[5] These maps are in the great tradition of medical cartography advanced by the work of Dr. Snow in the nineteenth century. Space prevents reference to other privately produced thematic sheet maps of particular interest to workers in many fields, but perusal of back numbers of appropriate journals will reveal the variety, quality, and extent of such coverage.

A very wide range of maps dealing with communications has appeared in recent years but many of these are known and used only by

---

"Towards the Improvement of the Dot Map," *The International Yearbook of Cartography,* vol. 7 (1967), 157–66; J. Ross MacKay, "Some Problems and Techniques of Isopleth Mapping," *Economic Geography,* vol. 27 (1951), 1–9; Philip W. Porter, "Putting the Isopleth in its Place," *Proceedings of the Minnesota Academy of Science,* vol. 25–26 (1957–58) 372–84; James J. Flannery, "The Graduated Circle: A Description, Analysis and Evaluation of a Quantitative Map Symbol," unpublished Ph.D. dissertation (University of Wisconsin, 1956).

Several major cartographic elements and symbols are considered in Arthur H. Robinson, *The Look of Maps: An Examination of Cartographic Design* (Madison: The University of Wisconsin Press, 1952); and Norman J. W. Thrower, "Relationship and Discordancy in Cartography," *The International Yearbook of Cartography,* vol. 6 (1966), 13–24.

[4] Calvin Schmid, *Handbook of Graphic Presentation* (New York: Ronald Press, 1954). This book, by a sociologist, suggests applications of a variety of graphic techniques including cartography to phenomena involving man particularly.

[5] Jacques May, "Medical Geography: Its Methods and Objectives," *The Geographical Review,* vol. 40 (1950), 9–41; one example in a series by the same author is, "Map of the World Distribution of Cholera," *The Geographical Review,* vol. 41 (1951), 272–73 plus insert map; and J. K. Wright, "Cartographic Considerations: A Proposed Atlas of Diseases," *The Geographical Review,* vol. 34 (1944) 649–52.

specialists.[6] On the other hand, transportation maps are among the most familiar of cartographic works. Maps used for public transportation (bus, train, underground, etc.) are often schematic or diagrammatic and emphasize only those features, such as stopping places, of immediate concern to the traveler. It should be mentioned in passing that the building of railways in a east-west direction over hundreds or even thousands of miles, as was done in the United States, underscored the necessity of the standardization of time zones; agreement upon this matter was reached at an international conference held in Washington, D.C., in 1884. Twenty-four zones of 15 degrees longitude each (the first being 7½°E and 7½°W of the Prime Meridian of Greenwich—0° longitude) theoretically bounded by meridians, were prescribed. Actually there are many deviations from the meridianal limits because of the desirability of political units—countries, states, provinces, and even in some instances counties—having uniform time throughout. When expressed cartographically, the relatively complicated time zone map results. Some airlines have been enterprising in providing interesting maps for their passengers. Outstanding in this regard is the work of Hal Shelton who prepares "natural color" maps of great aesthetic appeal. Shelton perfected this method of relief representation in which color, rather than expressing altitude per se, is used to indicate the expected hue of the surface (vegetation, etc.) at the height of the growing season. This scheme, combined with shading, gives a remarkably realistic picture of the earth as it appears to the air traveler and Shelton's maps have been used extensively by airlines. Such maps often have an overprint of the approximate route of the plane.

Perhaps the most familiar sheet map to the general public, especially in the United States, is the automobile road map. In this country over 200 million such maps are now distributed annually, mainly by oil companies. The ancestry of the road map can be traced to the earliest cartographic efforts, but the modern automobile map is a product of this century. State highway commissions and automobile associations are significant producers of road maps, and in some countries tire companies are the chief purveyors of automobile maps. But in terms of numbers of road maps published in the United States, the oil companies have dominated this activity since they began free distribution in 1919. The auto-

6 Gerald L. Greenberg, "A Cartographic Analysis of Telecommunication Maps," unpublished M. A. thesis (University of California, Los Angeles, 1963). This work contains a survey and analysis of over forty types of maps used in telecommunications systems and services; they are considered in respect to development, sources, design, scale, projection, etc. Among the maps discussed and illustrated in this work are radio, telephone, communications, Loran, radar, network, and engineering charts.

mobile map superseded the verbal guide as roads were improved and markers erected. In the early 1920's various states experimented with road numbering methods and the federal numbered highway system was approved by Congress in 1926. During this period the symbolization, as we know it today, was being developed by the three private map companies who still control most of this business in the United States. It is claimed that the modern road map probably presents more information per unit of area than any other common form of printed matter.[7]

Although the basic symbolization regarding road classification, etc., was developed in the 1920's, there have been many improvements in the design of automobile maps since that time. Color printing and plastic scribing techniques, as well as preprinted lettering and patterns, have been used with conspicuous success in road map production. Figure 9.2 shows the highway symbols recommended by the American Association of State Highway Officials and their use on a map of part of the Chicago area. These recommendations, however, have not always been followed by the map publishers, and road maps vary considerably one from the other. Not shown on this sample are such features of the road map as lists of streets with map references and coordinate systems for these, tables of distances between places on the map, and large scale and small scale inset maps. The modern American automobile map, because of constant revision, is normally the most current source of information on roads, boundaries, places of interest, etc., of the area covered. However, road maps characteristically lack relief representation and for this and other reasons, do not take the place of topographic maps as a general source of information concerning the landscape.

Just as routes have been a feature of maps since the beginning of cartography, so has terrain representation. We have seen how, during most of cartographic history, hills and mountains have been represented in profile or three-quarter view in contrast to the plan view of the maps on which they appear. We have discussed, also, the relatively late development of planimetrically correct methods of terrain rendering such as the quantitative isobath and contour and the qualitative hachures and shading. Physiographic diagrams and landform maps, which became popular in America in the mid-twentieth century, represent a return, geometrically, to the earlier period of cartography—even though in their richness of information the best examples are a great advance over earlier efforts at non-planimetrically correct terrain representation. The

[7] Roderick C. McKenzie, "The Development of Automobile Road Guides in the United States," unpublished M. A. thesis (University of California, Los Angeles, 1963).

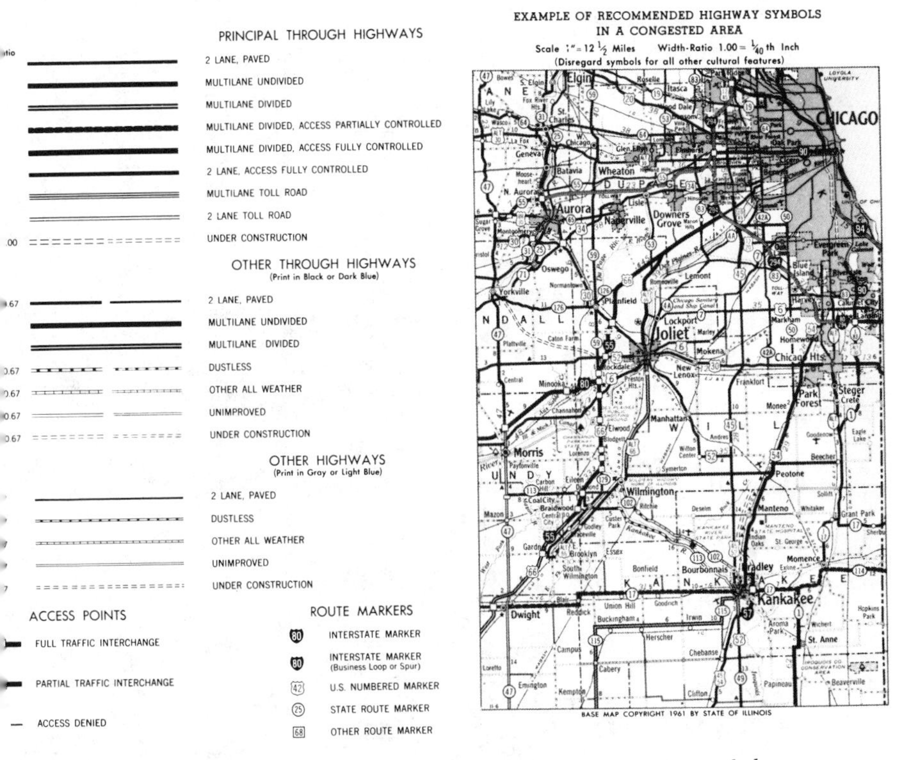

FIG. 9.2. *Conventional symbols for highway maps recommended by the American Association of State Highway Officials.*

physiographic diagrams of Armin K. Lobeck, who refused to use the term map for these renderings because of the inconsistency of the view of the terrain in relation to the map base, and the landform maps of Erwin J. Raisz and others were actually developed from the block diagram technique.

Followers of William Morris Davis in this century took the surface symbols of the block diagram and placed them on a map base. The result is a useful and easily understood picture of an area (Fig. 9.3), but the inconsistency of the perspective landform drawing and the plan view of the map has been a source of concern to a number of cartographers. The basic problem, of course, is that the landform symbols are not planimetrically correct, which becomes progressively more of a problem as the scale of the map increases. In an effort to illustrate the brightness of the configuration of a surface, Kitirô Tanaka, a Japanese cartographer/engineer, developed and utilized traces of parallel inclined

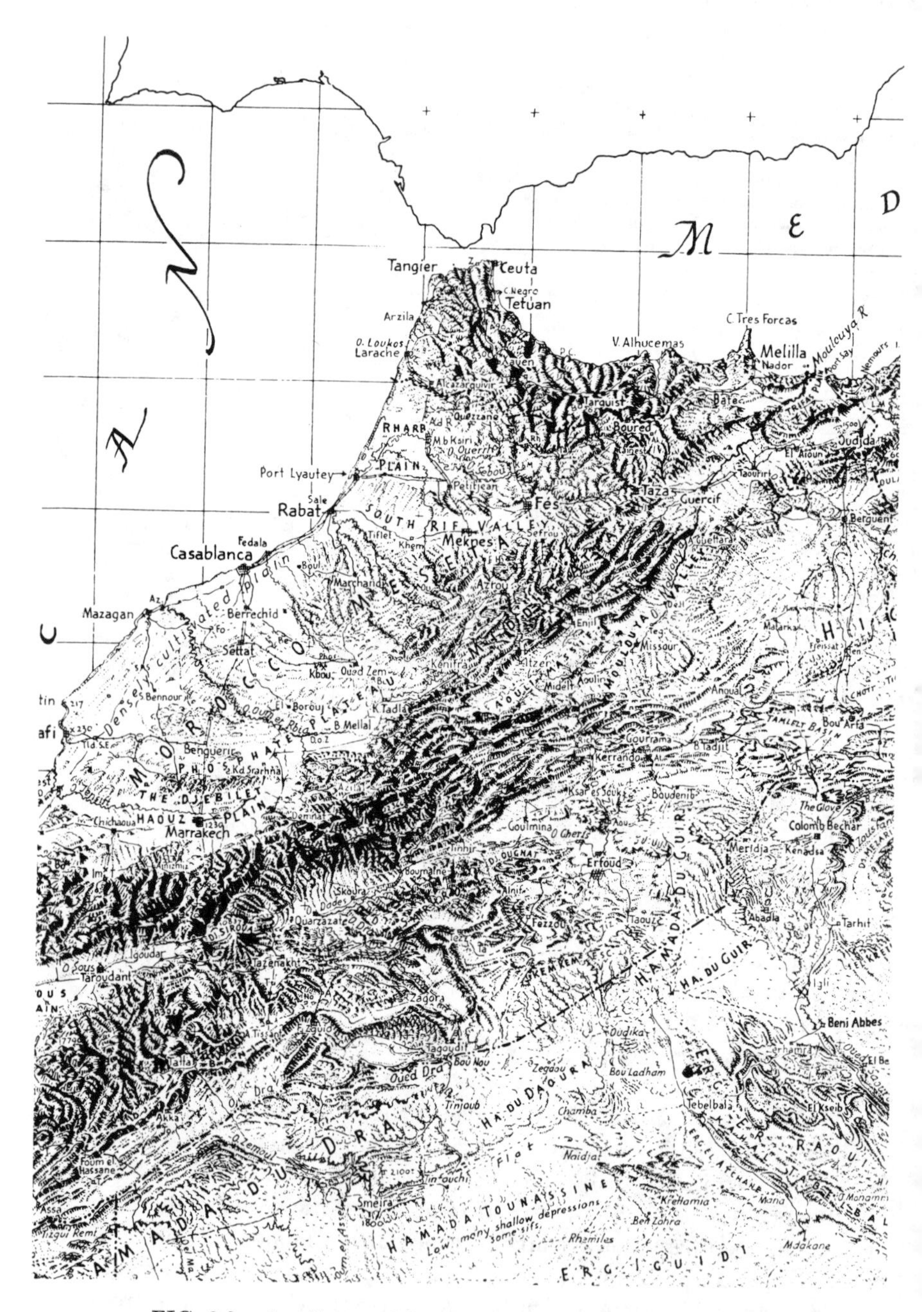

FIG. 9.3. *Small part of Landform Map of North Africa, by Erwin J. Raisz.*

planes, or, as he called them, "inclined contours."[8] By a rather simple technique, Tanaka transformed horizontal contours to the traces of the intersection of parallel, inclined planes with a surface. The traces used by Tanaka resulted from planes being inclined at 45 degrees; Figure 9.4 shows the relationship between the horizontal contour, the profile, and the trace of an inclined plane. Maps drawn with a series of traces of parallel planes inclined at 45 degrees, as recognized by Tanaka and others who commented on the technique, have a number of objectionable features. Among these are: (*1*) landforms appearing too flat; (*2*) an overall darkness to the rendering, which makes overprinting difficult, and (*3*) a pseudoscopic effect. For these reasons Tanaka's method remained virtually unused for a quarter of a century. In the mid-1950's Arthur H. Robinson and the author experimented with traces of planes having angles other than 45 degrees with the horizontal.[9] They used the traces to define forms and discovered that whatever the angle used, correct planimetry was retained. By using angles of less than 45 degrees with the datum, objections *1* and *2* above were overcome, while *3* was resolved by drawing fewer traces on the side facing a hypothetical light source or, conversely, heavier lines on the side away from this light source. The result is a planimetrically correct method of terrain rendering which permits the preparation of effective landform type maps of a larger scale than were possible previously, and which can be overprinted with information taken

[8] Kitirô Tanaka, "The Relief Contour Method of Representing Topography on Maps," *The Geographical Review*, vol. 40 (1950), 444–56.
[9] Arthur H. Robinson and Norman J. W. Thrower, "A New Method of Terrain Representation," *The Geographical Review*, vol. 47, no. 4 (1957), 507–20.

FIG. 9.4. *Diagram illustrating in two- and simulated three-dimensional forms the relationship between: a contour (x); a trace of an inclined plane (y); and a profile (z).*

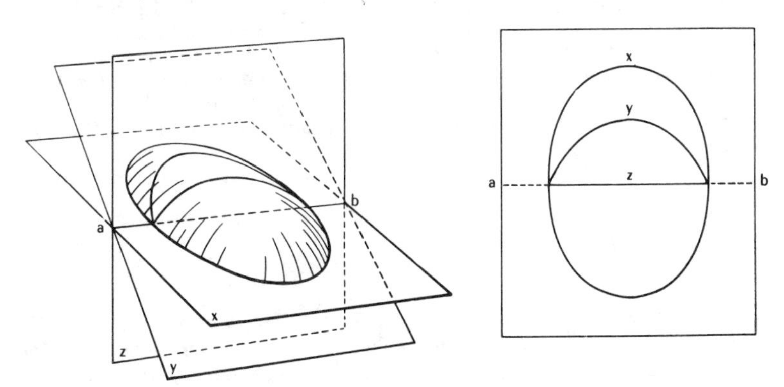

directly from vertical air photos or other maps (Fig. 9.5). Since the appearance of the Robinson and Thrower studies, the method described above has been adopted by planners, geologists, geomorphologists, and other specialists in many parts of the world.

The appearance of the third dimension in cartography has been achieved in various ways in addition to those discussed above. We have illustrated shaded relief on a topographic map; this technique has been used in a remarkably effective way at geographical scale in maps made by Richard E. Harrison.[10] Swiss cartographers, notably Eduard Imhof in recent years, have produced splendid terrain representation employing shaded relief and aerial perspective, in which warmer colors are used for higher elevations and cooler ones for lower areas.[11] Photographic and contour line anaglyphs (composite stereoscopic images) have been employed with great success in the French *Relief Form Atlas.*[12] Though hardly classified as maps, a large variety of raised (tactile) globes, parts of globes, and terrain models are now available. At one time these were usually made of plaster, often painted in realistic colors, but now other materials, particularly plastic, are used. These new materials have made the raised model easier to manufacture and lighter, but have not solved the great storage problem associated with such visual aids—many maps can be stored in the space required for a single model. One of the major problems with raised models, and indeed with profiles, is the degree to which these devices should be exaggerated in the vertical dimension to

[10] Richard E. Harrison, *Look at the World: The Fortune Atlas for World Strategy* (New York: Knopf, 1944). This is an atlas of perspective type terrain renderings, many on orthographic projections, showing the globe from unusual orientations, and is based on Harrison's work for *Fortune* magazine during World War II. Harrison now usually draws shaded relief in plan view; the most readily available source of this form of his rendering is found in, Norman J. W. Thrower, ed., *Man's Domain: A Thematic Atlas of the World,* 2nd ed. (New York: McGraw-Hill, 1970).

[11] Eduard Imhof, *Schweizerischer Mittelschulatlas* (Zürich: Konferenz der Kantonalen Erziehungsdirektoren, 1963); this popular European atlas is the practical result of the author's lifelong concern with terrain representation, the theoretical basis of which is contained in his *Kartographische Geländedarstellung* (Berlin: Walter de Guyter and Co., 1965). John S. Keates, "The Perception of Colour in Cartography," *Proceedings of the Cartographic Symposium* (Edinburgh, 1962), pp. 19–28.

[12] A. Cholley, et al., *Relief Form Atlas* (Paris: Institut Géographique National, 1956).

FIG. 9.5. (opposite) *Above, central portion of a relief map of part of the Colorado Rockies using traces of inclined planes as a basis. Scale of original 1:50,000, greatly reduced. Below, large-scale section of the map above, with overlay of selected surface information.*

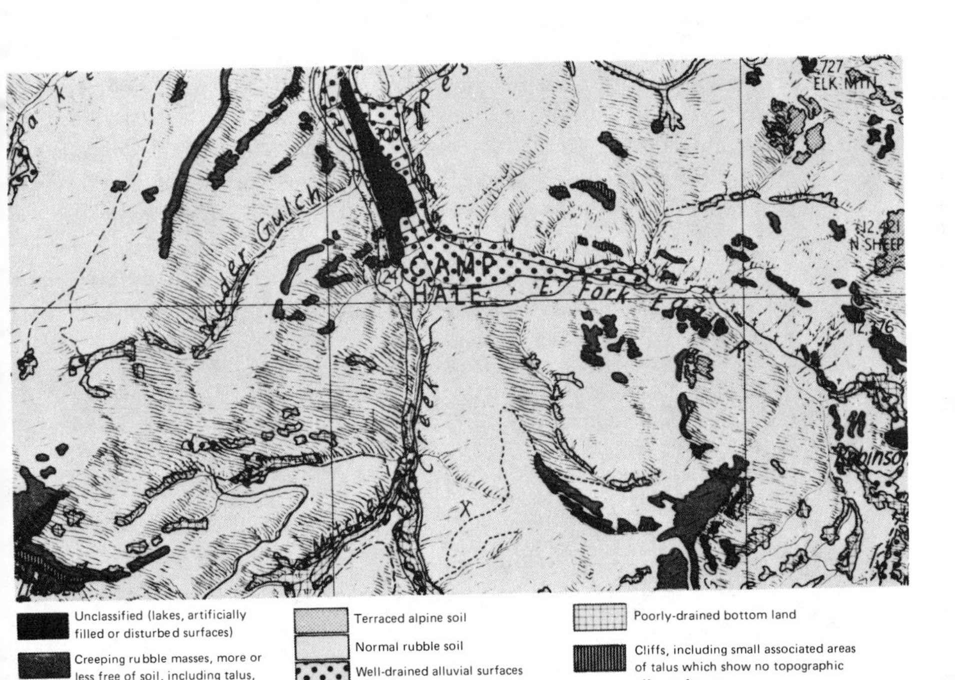

Because terrain types intergrade and mingle, boundaries on this map separate areas of dominance by the various types shown, but are not always occupied by such types exclusively.

give a realistic appearance.[13] Models, profiles, etc., tend to look too flat unless vertically exaggerated because man is apt to think of elevations on the land as being relatively higher than they are in reality. A tactile surface has a very special use in maps for the blind which have been experimented with in recent years.[14]

The third dimension, whether simulated or actual, is not confined to representation of landforms. We have illustrated statistical surfaces in two- and simulated three-dimensional form (Fig. 7.8). The traces of inclined planes have also been used for population, rainfall, and other distributions.[15] Isometric maps of cities in the tradition of those of the Renaissance (Fig. 5.7) are now being made by, among others, Hermann Bollmann, a German cartographer, and his associates. After making bird's-eye view maps of a number of European cities, Bollmann turned his attention to New York. His representation of that city must be among the most remarkable maps of modern times showing, as it does, buildings and other features with great accuracy and detail (Fig. 9.6). Artists such as Bollmann, Shelton, and Harrison are producing maps today that are artistically the equal of the best work of cartographers of the past. Furthermore, the techniques by which their maps are reproduced are vastly superior to earlier methods which were incapable of bringing out all of the subtlety of original artwork.

Special uses of maps include the representation of countries on postage stamps and maps as cartoons; in the latter, the outlines of the country have often suggested an appropriate figure, etc. Maps reproduced in magazines and newspapers reach a very large number of people who might otherwise be little concerned with cartography. Such maps are so well-known that it is unnecessary to discuss them at any length or to reproduce examples. They range from sophisticated and highly original maps in color for magazines printed on high grade paper, to simple line drawings in black and white which will reproduce satisfactorily on soft

[13] George F. Jenks and Michael R. C. Coulson, "Class Intervals for Statistical Maps," *The International Yearbook of Cartography,* vol. 3 (1963), 119–34; George F. Jenks and Fred C. Caspall, *Vertical Exaggeration in Three-Dimensional Mapping,* Technical Report no. 2, NR 389–146 (Lawrence: Dept. Geography, University of Kansas, 1967); George F. Jenks and Paul V. Crawford, *Viewing Points for Three-Dimensional Maps,* Technical Report no. 3, NR 389–146, *ibid.*

[14] John C. Sherman and Willis R. Heath, "Problems in Design and Production of Maps for the Blind," *Second International Cartographic Conference,* Series II, no. 3 (1959), 52–59.

[15] Arthur H. Robinson, "The Cartographic Representation of the Statistical Surface," *The International Yearbook of Cartography,* vol. 1 (1961), 53–63; Norman J. W. Thrower, "Extended Uses of the Method of Orthogonal Mapping of Traces of Parallel, Inclined Planes with a Surface, especially Terrain," *The International Yearbook of Cartography,* vol. 3 (1963), 26–38.

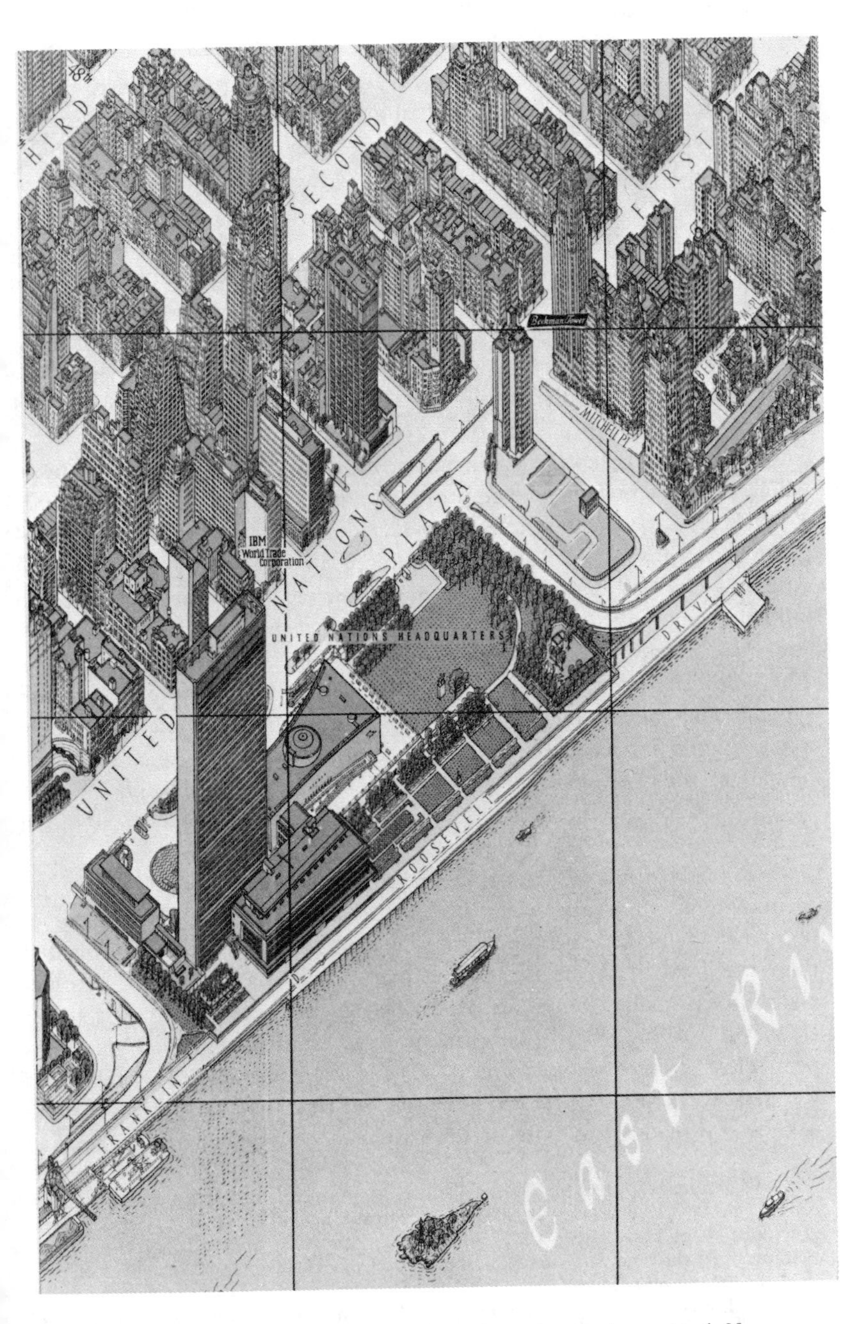

FIG. 9.6. *Slightly reduced section of an isometric map of New York City focusing on the United Nations Plaza, by Hermann Bollmann.*

newsprint.[16] In the former category, are the maps of Harrison for *Fortune* magazine. Like those of Shelton, these maps are interesting, from a technical point of view, as examples of cartography where the color separation is normally accomplished by the photographer from the finished artwork (four color process) rather than by the cartographer, as is usual with topographic and most other colored maps (flat color). Maps showing parts of the world from unfamiliar viewpoints by Harrison, and those of the Pacific area by the Mexican artist, Miguel Covarrubias, originally published in *Fortune,* have become classic examples of their respective cartographic types. At the other end of the scale are newspaper maps which are usually simple line drawings reproduced in black and white.

It is a short step from journalistic cartography to propaganda maps, which are designed to give impressions and in which the map becomes a psychological tool. A whole arsenal of devices may be used for propaganda mapping: color; projection; maps in series; size relationships; symbols—especially arrows; and boundaries; etc. These are employed to influence opinion rather than to dispassionately inform the map reader.[17] Advertising maps are also used to influence, but for a different purpose. The authority of the map and globe which is emblematic of, and synomomous with, education is invoked, even though some advertising maps contain intolerable errors. For example, the author once saw such a map of the Americas in which the Equator passed through Panama rather than Brazil and Ecuador. We should be properly critical of all printed materials—including maps!

Publicly displayed maps are a familiar sight in various cities and towns, at bus and railway stations, etc.[18] Sometimes they are made of permanent materials such as glazed tiles, but more often they are printed maps enclosed in a case. Often the map is pictorial in character, especially if of a smaller tourist center. In recent years guide maps of a rather elaborate type have been placed in strategic locations, especially in European cities, housed in substantial, glass-fronted installations. Frequently the map is mounted on cloth and attached to rollers so that by turning a knob the reader can raise or lower it to a desired height. A classified

[16] Walter W. Ristow, "Journalistic Cartography," *Surveying and Mapping,* vol. 17, no. 4 (1957), 369–90; James F. Horrabin, *An Atlas of Current Affairs* (London: Victor Gollancz Ltd., 1934). The maps in this atlas are excellent examples of simple journalistic political maps from between the Wars by a master of this cartographic form.

[17] Louis O. Quam, "The Use of Maps in Propaganda," *The Journal of Geography,* vol. 42, no. 1 (1943), 21–32. In addition to propaganda maps, there are a number of maps which are obviously designed to mislead for purposes of fraud.

[18] Norman J. W. Thrower, "The City Map and Guide Installation," *Surveying and Mapping,* vol. 22, no. 4 (1962), 597–98.

register of places of interest to the tourist as well as alphabetical lists of streets with map references are typical features of such display maps. The permanently installed, vertically mounted map is a device of utility to visitors to cities but special types, some of which are electronically controlled to display needed information to businessmen, clients, etc., are also used in offices.

The subjective map and the map which consciously distorts from an areally correct shape are two cartographic forms which have been increasingly used in recent years. Of course the possibilities for subjective maps are limitless. They range all the way from interpretations of individual perceptions of phenomena which may be only in the mind and, if expressed in some manner, may or may not be intelligible to anyone but the originator, to such well-known examples as "The New Yorker's Idea of the United States of America," the humor of which can be appreciated by millions of people.[19] In this map the sizes of places in the United States are in relation to their assumed importance to the inhabitant of New York City, so that Manhattan is considerably larger than the State of Washington (which, incidentally, is located south of Oregon). The size of an area, however, is not simply a function of distance from New York, for Florida appears larger than all the states intervening between it and New York. The so-called APT (Area Proportional To) principle also has been used for serious purposes, as in Pierre George's well-known cartogram showing the size of countries of the world according to population.[20] The same concept has been used for a map "The Political Colour of Britain by Numbers of Voters" published in *The Times,* London (October 19, 1964) in which "(*1*) areas of the map are proportional to the population and not to the amount of land, (*2*) contiguity of areas is everywhere preserved, and (*3*) the relative disposition of places on the map is as close as possible to the relative geographic location" (p. 18). A quarter of the bulk of the map is in the populous southeast, around London, while the north of Scotland is much smaller than its considerable geographical size would suggest. To illustrate this principle we reproduce three maps of Hungary adapted from a study of economic maps by Lászlo Lackó (Fig. 9.7).[21] The maps

19 J. Russell Smith and M. Ogden Phillips, *North America* (New York: Harcourt, Brace and Company, 1949), p. 169. The map by Daniel K. Wallingford is reproduced in this well-known geography text. It was republished, along with other maps in which cartographic geometry has been modified in Waldo R. Tobler, "Geographic Area and Map Projections," *The Geographical Review,* vol. 53 (1963), 59–78. See also John E. Dornbach, "The Mental Map," *Annals of the Association of American Geographers,* vol. 49 (1959), 179–80 (abstract).

20 Pierre George, *Introduction a L'Étude Géographique de la Population du Monde* (Paris: L'Institut National D'Études Demographiques, 1951).

21 Lászlo Lackó, "The Form and Contents of Economic Maps," *Tijdschrift Voor Econ. En Soc. Geografie,* vol. 58 (1967), 324–30.

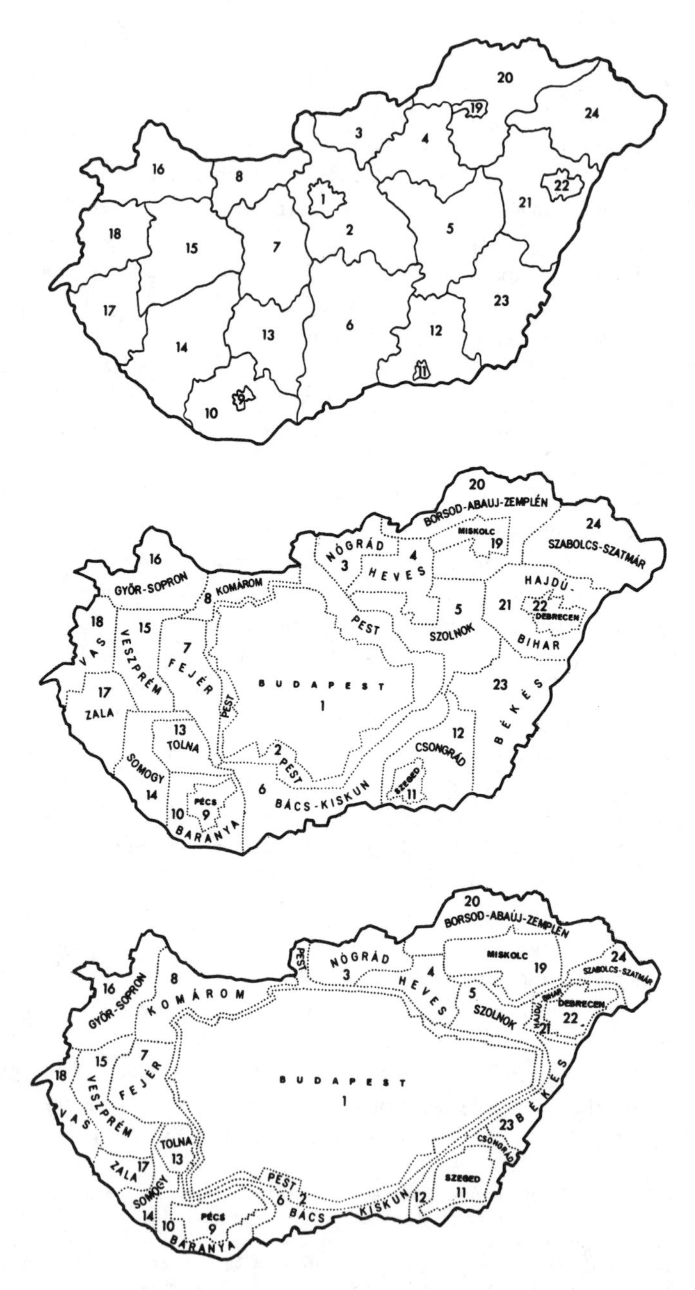

FIG. 9.7. *Three maps showing the size of the counties of Hungary (from top to bottom): according to geographical area; in proportion to population; and in proportion to the number of industrial employees.*

are all drawn within a framework provided by the geographical limit of the country, but the area of the internal political units differs proportionally according to the phenomena being mapped.

Area is not the only geographical quality or element which can be altered to suit particular purposes. For example, on a much-reproduced map by Torsten Hägerstrand distances from the parish of Asby are scaled logarithmically to provide a more useful cartographic base for plotting migration flow, telephone calls, and other distributions.[22] Among the European cartographers the Scandinavians, particularly Swedes, enjoy a reputation for originality in statistical mapping equal to the Swiss in terrain representation. The economic maps of W. William-Olsson, for example, are extremely informative and contain a number of cartographic innovations. Problems associated with economic mapping over large areas of Europe are much greater than in the United States because of the lack of uniformity in the census reports of the several countries involved. Nevertheless, attempts have been made to reduce the information to a common base, as in Olsson's *Economic Map of Europe* of the scale of 1:3,250,000 and in the *Atlas of Western Europe* by Jean Dollfus.[23] Recently an attempt has been made on the part of a private American map publisher to produce a thoroughly international atlas of the world.[24] This was developed in co-operation with similar cartographic organizations in Hungary, Sweden, Britain, Germany, and Japan. A very valuable feature of this atlas is the consistent use of a few scales for certain classes of maps, including plans of all major metropolitan areas of the world on the scale of 1:300,000. A problem which almost defies solution in such a work is the rendering of place names on the maps.

We have seen how through the centuries man has solved the problem of transferring the spherical grid (graticule) from a globe to a plane surface. We have also noted that, although all global properties cannot be preserved in any one projection, these devices may possess positive qualities that make them more than poor substitutes for the globe. Recent advances have been made to this study by the development of new projections, adaptations or new cases of existing projections, and new uses for projections devised in earlier times. Relatively few individuals have invented original, useful projections but all cartographers should know enough about projections to make good choices among those that exist. Indeed the nonspecialist who looks at maps intelligently

22 Tobler, "Geographic Area," 65.

23 Jean Dollfus, *Atlas of Western Europe* (Chicago: Rand McNally and Company, 1963).

24 Robert L. Forstall et al., eds., *The International Atlas* (Chicago: Rand McNally and Company, 1969).

ought to know something of the strengths and limitations of the projections used (Appendix A).

Various attempts have been made to classify projections, one profitable way being in terms of geometrical shapes which may actually or theoretically be used for their construction.[25] Many shapes may be employed for this purpose; three commonly used being a cylinder, a cone, and a flat surface giving rise, respectively, to cylindrical, conical, and azimuthal classes of projections (Fig. 9.8).[26] Thus one may have: a cylinder touching a generating globe along one line or intersecting to touch along two lines, therefore spreading the distortion or deformation (such a line of contact is known as a standard line and is often, but not necessarily, a parallel); a cone touching along one line or intersecting to touch along two lines; a flat surface in contact at one point, or intersecting, or at a distance from the generating globe. One may further conceive of a transparent globe (or hemisphere) on which are drawn opaque lines of latitude and longitude and a light source with which to "project" the grid onto the shapes described above to produce projections with different qualities. For example, a light source at the center of the globe projecting onto a flat surface in contact with globe at one point, produces a Gnomonic projection; if the light source is on the side of the globe exactly opposite to the flat surface, a Stereographic projection results; whereas if the light source is at an "infinite" distance, the Orthographic projection is produced. All of these projections, devised centuries ago, have found new and important uses in the modern world of rapid transportation, communication and global thinking.

As an example of a projection devised recently, we may call attention to the Dymaxion Airocean World Projection devised by R. Buckminster Fuller, the inventor of the geodesic dome.[27] Fuller's projection consists of twenty equilateral triangles, the boundaries of which are great circles. The projection can be bent on its triangular edges to form a planar faceted icosahedron, or the triangles can be displayed together on a plane surface hinged along different boundaries to show interesting earth relationships. The sizes and shapes of earth areas closely approximate those on the globe. A projection devised by Osborn M. Miller of

[25] Most map projections are prepared from tables originally, but they may be transformed from one scale or form to another optically or graphically, or by use of computer plotters.

[26] Only certain shapes are likely to be useful. In addition to the regular figures tangential to the globe, i.e., tetra-, hexa-, octa-, dodeca-, and icosa-hedron, other forms have been used. These include toroidal (doughnut shape), paraboloidal, hyperboloidal, pyramidal, and cataclysmal!

[27] R. Buckminster Fuller, "Dymaxion Airocean World Map," with descriptive notes (Raleigh: Student Publications of the School of Design, North Carolina State College, 1954).

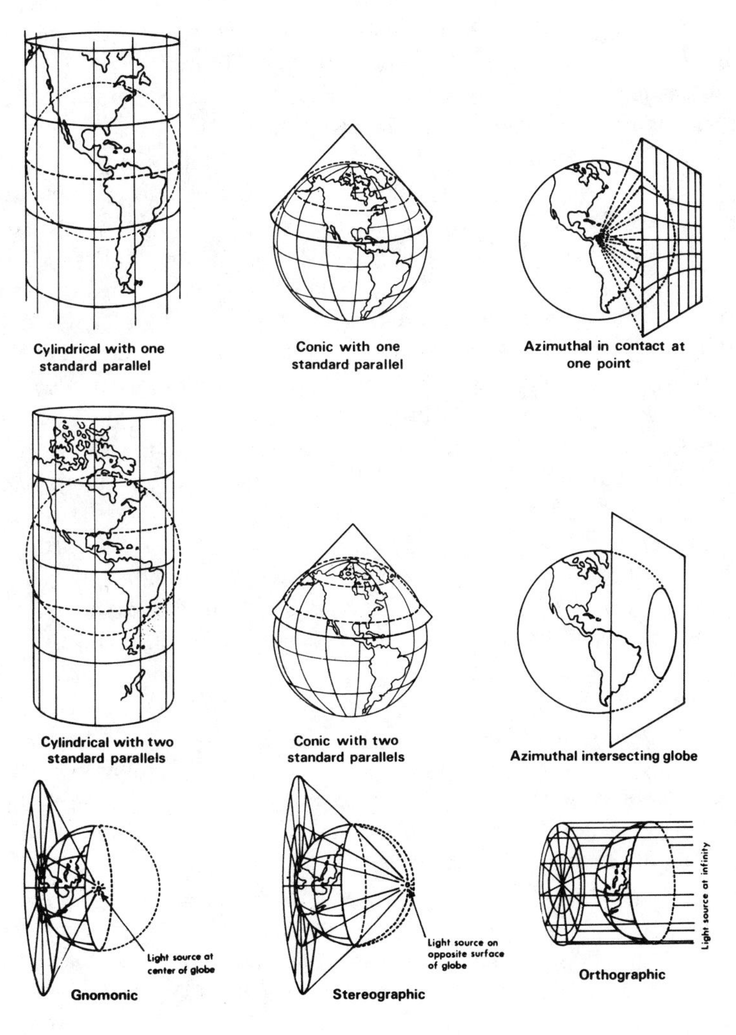

FIG. 9.8. *Geometrical properties of various cylindrical, conic, and azimuthal projections.*

the American Geographical Society and named after its inventor, can be cited as another example of a projection developed in recent years. The Miller Cylindrical Projection forms a rectangle, as does the Mercator, but unlike that projection, allows the poles to be shown. It avoids the extreme deformation of the Mercator in high latitudes and the great angular exaggeration of the Cylindrical Equal Area projection and, therefore, is a useful compromise for displaying the whole earth.[28]

In 1923, J. Paul Goode of the University of Chicago grafted parts of two existing equal-area projections to form the so-called Homolosine; he combined the lower latitude section of the Sinusoidal (latitude 0–40 degrees N and S) with the higher latitude parts of the Mollweide (latitude 40–90 degrees N and S). An equal-area projection which is the arithmetical mean between the Mollweide and the Sinusoidal was developed by S. Whittemore Boggs in 1929. Early in his career Goode produced an interrupted form of the Mollweide (Homolographic) and he later applied the same principle to his Homolosine projection. We have seen the quality of interruption, a technique for improving the shape of the representation, in the gores of the globe and alluded to the (interrupted) Double Cordiform Projection of Mercator. In his interrupted version of the Homolosine, Goode used a series of lobes, each with a standard meridian but hinged along a common equator. A further modification of the grid involves condensation in which areas of little interest are deleted and the remaining areas are moved together to save space. In all maps, whether simple or complex, an indication of the grid either by continuous lines or by crosses and ticks should be given so that the geometry of the projection used may be understood.[29]

Oblique and transverse cases of well-known projections have been used increasingly in recent times. In the transverse (polar) case the grid is shifted 90 degrees so that, for example, the Transverse Mercator is tangent upon a meridian rather than the equator as it is in its standard or conventional form. If the shift is less than 90 degrees from the normal position, the term oblique is used. In its transverse and oblique cases, the Mercator remains conformal (grid lines cross at right angles and shapes around a point are correct) but the quality that all straight lines are rhumb lines (lines of constant compass direction) is not retained. Although any projection can be shifted in the manner described above,

[28] Irving Fisher and Osborn M. Miller, *World Maps and Globes* (New York: Essential Books, 1944), esp. p. 79; and Osborn M. Miller, "Notes on Cylindrical World Map Projections," *The Geographical Review,* vol. 32 (1942), 424–30.

[29] Richard E. Dahlberg, "Maps Without Projections," *Journal of Geography,* vol. 60 (1961), 213–18; and Richard E. Dahlberg, "Evolution of Interrupted Map Projections," *The International Yearbook of Cartography,* vol. 2 (1962), 36–54.

in practice, only a few are normally so treated including, among equal-area projections the Mollweide, and among conformal ones the Mercator. Oblique projections are often used for world maps in modern atlases, while transverse projections, especially the Mercator, have gained increasing popularity for the topographic map series of several of the great surveys of the world (including Britain's Ordnance Survey). Those who have contributed to the development of map projections include mathematicians, government and nongovernment cartographers, and nonprofessionals. Among the latter group, interestingly, a number of clergymen have been prominent.

An idea having to do with map projections that deserves more attention than it has received is the grid formed by meridians and parallels for the comparison and measurement of area, proposed by Harry P. Bailey.[30] Bailey employs meridians and parallels to delimit areas of equal size on the surface of the earth. To accomplish this, meridians remain at a constant and equal angular interval while parallels are spaced in such a manner that quadrilaterals of uniform size on the globe are enclosed. This has the effect of producing an area-reference grid of considerable utility. Bailey has applied this principle to a variety of projections—equal area, conformal, and compromise types.

Today cartography is moving in a number of new directions.[31] One such direction is computer cartography. The computer is generally capable of storing more information than a map, but complex spatial associations are not evident unless they are displayed cartographically. Such information as area, distance, direction, contiguity, and relationship can be computerized and a number of standard programs for computer mapping are now available. One such program is the Synagraphic Mapping System (SYMAP) which is capable of composing spatially distributed data in map, graph, and other forms of visual display.[32] In making a computer map the following simple steps are normally taken:

30 Harry P. Bailey, "A Grid Formed of Meridians and Parallels for the Comparison and Measurement of Area," *The Geographical Review,* vol. 46, no. 2 (1956), 239–45; and by the same author, "Two Grid Systems that Divide the Entire Surface of the Earth into Quadrilaterals of Equal Area," *Transactions of the American Geophysical Union,* vol. 37, no. 5 (October 1956), 628–35.

31 John C. Sherman, "New Horizons in Cartography: Functions, Automation and Presentation," *The International Yearbook of Cartography,* vol. 1 (1961), 13–19.

32 The SYMAP Project is administered through The Laboratory for Computer Graphics of Harvard University. Other methods, such as the Oxford Mark I System, are described in W. G. V. Balchin and Alice M. Coleman, "Cartography and Computers," *The Cartographer,* vol. 4, no. 2 (1967), 120–27.

For earlier uses of the computer in mapping, see William Warntz, "A New Map of the Surface of Population Potentials for the United States, 1960," *The Geographical Review,* vol. 54 (1964), 170–84; and Waldo R. Tobler, "Automation and Cartography," *The Geographical Review,* vol. 49 (1959), 526–34.

(*1*) coordinates of the controlling points are established on a manual digitizing board; (*2*) the information is entered on special coding forms; (*3*) information from the coding forms is transferred to punch cards; (*4*) the deck of cards is then fed to, and processed by, the computer (an operation which usually takes only seconds); (*5*) the map is printed out line by line at rates such as forty lines per second—an average printout taking less than a minute. Use of tapes and disks greatly facilitates the manipulation of large data banks, and, in any case, the whole operation may be completed in only a fraction of the time formerly required to produce similar results. The low cost and speed of creating these maps, made without benefit of hands, are valuable facets of computer cartography, as are its flexibility and objectivity. The ability to make different maps with ease from a data set and to compare them with each other and with those made from other data sets is obviously a great advantage in understanding the complexity of areal distributions. Furthermore, the reliability of the original data can be evaluated by means of computer mapping. The computer is capable of producing maps according to contour (isoline), choropleth (conformant) and point (proximal) methods (Fig. 9.9). Simulated three-dimensional representations can be drawn by the computer with coordinates supplied in plan or elevation form. Similarly map projections can be transformed from one case to another on a computer. Maps in series, e.g., population change in an area through time, can be composed into true, three-dimensional models. In fact, the whole problem of understanding surfaces, whether real or abstract, concrete or imaginary, is greatly facilitated by the computer. In spite of its present high degree of usefulness, there are certain areas in which computer mapping can be significantly improved. The symbolization normally used, which often consists of printing and overprinting of characters not designed specifically for mapping, is less than satisfactory visually. This is now being improved so that more logical value and textural progressions are attainable.[33] The "straight line" character of "contours" and other curving lines, in detail, owing to the fact that the computer plots from point to point is a further objection, which can be minimized by photographic reduction and other means. Standard output paper is only fifteen inches wide but this problem can be partially overcome by the computer automatically breaking the map into bands which the user can tape together. These are relatively minor problems and it must be recognized that the computer is transforming cartography as much as other technological developments, e.g., printing and aerial photography, changed the course of map-making at earlier periods. At present the computer map is more advanced in its information than in its design.

[33] Jacques Bertin, *Sémiologie Geographique* (Paris: Mouton, 1967).

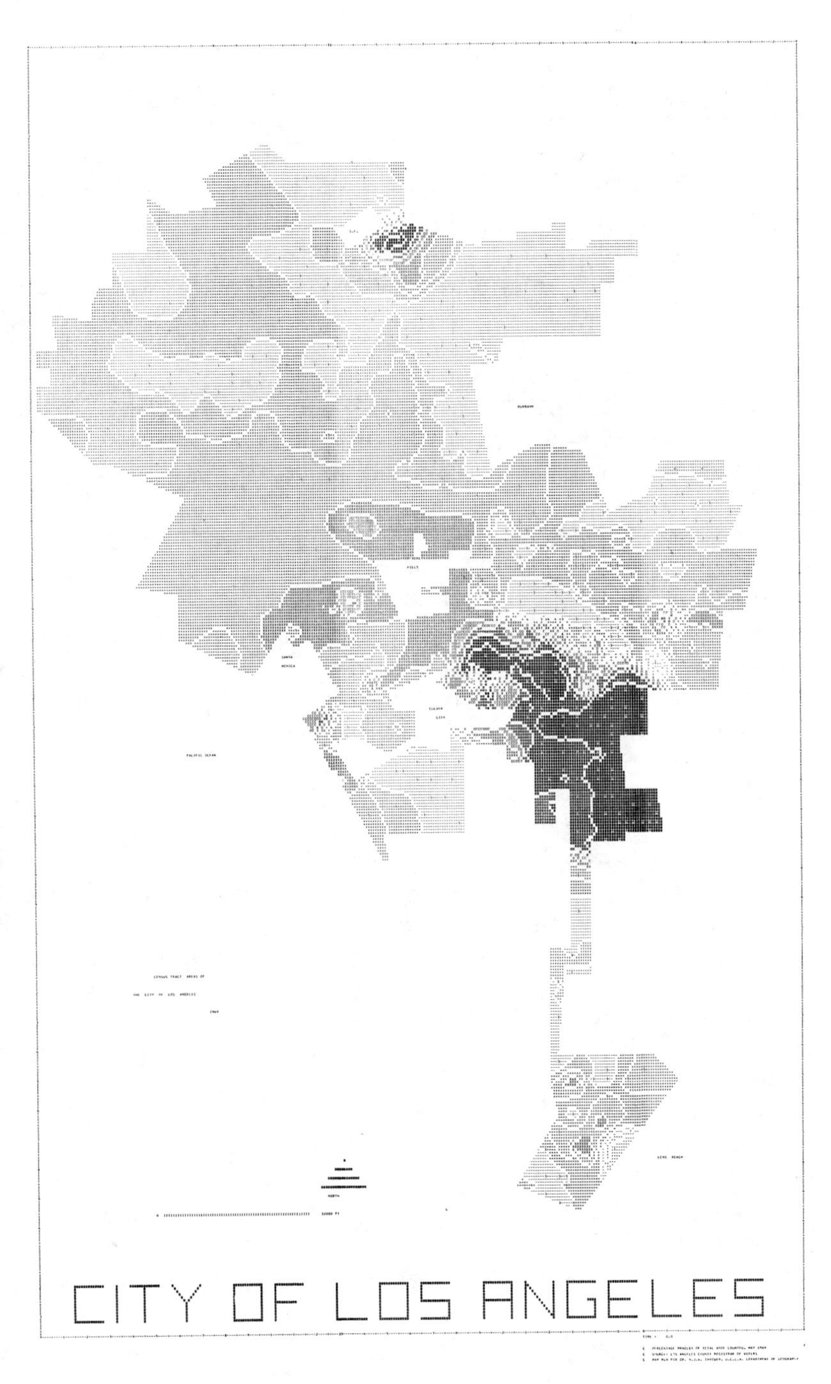

FIG. 9.9. *Computer map, greatly reduced, showing voting patterns in the City of Los Angeles in the 1969 mayoral election.*

However, experiments in computer "art" suggest that effects are possible, some of which have cartographic applications that are not attainable by other means. Already the use of computer driven pen plotters and cathode ray tubes give great promise for the solution of many technical problems in computer cartography.

We have dealt with the map as a snapshot representing phenomena at a given time. We have also suggested that a series of maps of the same phenomena at different times can be used to suggest temporal change. Furthermore, in the previous chapter it was indicated that experimental time-lapse movies have been made by putting a series of 6-hourly weather maps on film. In this and similar presentations the hiatus is short enough so that a mental idea of the preceding image is retained, but long enough so that no real impression of movement results. A true motion picture differs from a time-lapse presentation in that, in the former, a continuous appearance of motion is achieved. This is accomplished by presenting the different images so rapidly that the eye is unable to detect the short interruptions between the successive frames of the film. The illusion of movement thus produced, known as the *phi phenomenon,* is familiar through its use in advertising signs as well as in motion pictures. In sound films sequential frames projected on the screen at the rate of twenty-four per second are synthesized by the eye and the subject appears to move.

This principle has been used in very significant ways in cartography adding the fourth dimension, time, to maps.[34] It is probable that now more people see maps projected on television and movie screens (even if these are not always animated) than any other type of map. Different mass media utilize maps for a variety of purposes including weather reports, news programs, and entertainment, and some of them are abysmally poor examples of cartography. However, the following remarks will focus on animated cartography in educational motion pictures. Without going into detail concerning such methods, it can be said that all of the major techniques of making motion pictures have a cartographic counterpart. Thus a "live" movie can be made of a person actually drawing a simple map; this has the pedagogical advantages of a blackboard drawing where the image develops before the viewer. A variation of this is to prepare a movie of information being added to an existing map as, for example, the route of an explorer. Another method is to create regular animated sequences in the manner of the cartoon using a series of overlays or "cels." The same background map can be used throughout

[34] Norman J. W. Thrower, "Animated Cartography in the United States," *International Yearbook of Cartography,* vol. 1 (1961), 20–30; and "Animated Cartography," *The Professional Geographer,* vol. 11, no. 6 (1959), 9–12.

a sequence but the transparent cels each bear a slightly different image. The cels are photographed one by one over the background to form the individual frames of a movie. When assembled in order and projected on the screen at the correct speed, an animated map results. Computers are now being utilized in the preparation of animated films, which reduces much of the laborious process of making such presentations.

Thousands of movies have been made in which some map sequences appear. Symbols used in animated cartography include: dots for locations; arrows for directions; lines for transportation, communications, and boundaries; pictorial symbols for agricultural or mineral products; areal symbols for population, forests, etc. Lettering, like the other symbols, can be added at the appropriate moment and removed when no longer needed, a very great advantage of animated cartography which eliminates cluttering of the map. A number of cartographic elements considered essential in static cartography are often omitted on animated maps. Owing to the variable size of the projected image, a verbal scale or representative fraction should not be used but a graphical scale which enlarges and reduces with the image is, of course, perfectly satisfactory. One drawback is that the animated map does not permit prolonged study unless the projector is halted; however, for illustrating the dynamics of certain areal relationships, it is unexcelled. Animated cartography can most effectively unite the temporal element of history with the areal view of geography.[35]

Philosophers and physicists, princes and presidents, poets and painters, physicians and priests, planners and psychologists, perceptionists and programmers have in some fashion been concerned with the map. Beginning before the written record, cartography runs as a thread through history. Maps have been made by so-called primitive as well as by sophisticated peoples. By no means have professional cartographers been the only substantial contributors to this art and science; rather, because of its eclectic and universal nature, it has drawn its practitioners from many fields. The special relationship of cartography with geography has given cartography focus, but it must not be thought of as the handmaiden of this or any particular discipline. It exists separately, even while being of interest to both the layman and the specialist. The explosion of knowledge in very recent times is reflected in cartographic diversity and output; it is estimated that over half of all the maps ever produced have been made since the beginning of World War II.

35 H. Clifford Darby, "Historical Geography," in H. P. R. Finberg, ed., *Approaches to History* (London: Routledge and Kegan Paul, 1962), pp. 127–56, esp. p. 139.

Many of man's most important achievements—from philosophical considerations on the nature of the earth to setting foot on the lunar surface—have had cartographic expression and, in turn, have been advanced by cartography. Over the centuries cartography has developed its own methodology and traditions. Sometimes these have had an inhibiting effect on progress but, in spite of a tendency toward conservatism, cartographers have generally been sensitive to change and kept abreast of philosophical, scientific, and technological progress. Geographically, the centers of action in cartography have usually been where science is currently flourishing. As in the past, ingenious solutions will undoubtedly be found to problems associated with the representation of spatial phenomena in the future while "knowledge," as Tennyson reminds us, continues to "grow from more to more."

APPENDIXES

Appendix A

Name of Projection	*Century*	*Inventor*	*Family/Form*[1]
Gnomonic (Horoscope)	5th B.C.	Thales?	Azimuthal
Orthographic (Analemma)	2nd B.C.	Hipparchus?	Azimuthal
Stereographic (Planisphere)	2nd B.C.	Hipparchus?	Azimuthal
Marinus	A.D. 1st	Marinus of Tyre	Cylindrical
Ptolemy (Simple Conic)	2nd	C. Ptolemy	Conic
Ptolemy (Second (Projection)	2nd	C. Ptolemy	Miscellaneous
Cordiform	16th	M. Waldseemüller et. al.	Miscellaneous
Werner (Equal (Area)	16th	J. Stabius and J. Werner	Miscellaneous
Mercator	16th	G. Mercator	Cylindrical
Conic (with Two Standard Parallels)	16th	G. Mercator	Conic
Sinusoidal (Sanson–Flamsteed)	17th	N. Sanson	Miscellaneous
Cylindrical Equal Area (Plane Chart)	18th	J. H. Lambert	Cylindrical
Conic Equal Area	18th	J. H. Lambert	Conic
Conic Conformal	18th	J. H. Lambert	Conic
Azimuthal Equal Area	18th	J. H. Lambert	Azimuthal
Transverse Mercator	18th	J. H. Lambert	Cylindrical
Bonne	18th	R. Bonne	Miscellaneous
Albers	19th	H. C. Albers	Conic
Mollweide (Homolographic)	19th	K. B. Mollweide	Miscellaneous
Polyconic	19th	F. Hassler	Miscellaneous
Goode (Homolosine)	20th	J. P. Goode	Miscellaneous
Miller Cylindrical	20th	O. M. Miller	Cylindrical
Fuller Dymaxion	20th	R. B. Fuller	Miscellaneous

[1] Some authorities recognize: pseudo conic, e.g. Polyconic; Bonne,

Selected Map Projections

Salient Characteristics	*Principal Uses*
Straight great circles; limited area	Astronomy; later, route plotting
Hemisphere—viewed from infinity	Astronomy; later, earth illustrations
Conformal, circles project circles	Astronomy; later, air navigation
Straight line meridians, parallels	Early maps of known world/regions
Straight meridians, curved parallels	Early maps of known world/regions
Curved meridians, curved parallels	Early maps of known world/regions
Similar to preceding, less distortion	Early maps of known world/regions
Equivalent; pole, standard parallel	Early maps of known world/regions
Straight loxodromes; conformal	Navigational charts; world maps
Less deformation than simple conic	Continents, countries, regions
Equivalent; meridians—sine curves	World distributions; atlas maps
Equivalent; straight meridians, parallels	World, countries, states, cities
Equivalent; one standard parallel	Mid-latitude maps, e.g. U.S.A.
Conformal, two standard parallels	Mid-latitude maps; air navigation
Equivalent, directions true from a point	Hemisphere, polar, and air line maps
Conformal; central meridian true	Topo., maps; illustrating air routes
Equivalent; parallels true to scale	Mid-latitude maps; e.g. U.S.A.
Equivalent; two standard parallels	Topo., maps; continents; e.g. Europe
Equivalent; elliptical meridians	World distributions—atlas maps
Parallels non-concentric, standard	Topo., maps; modified—air charts
Composite projection; interrupted	World distributions—atlas maps
Minimizes shape, size deformation	World distributions—atlas maps
Great circle bounded icosahedron	Illustrating global relationships

pseudo cylindric e.g. Sinusoidal; Mollweide; Goode.

Appendix B

Short List of Isograms[1]

General: Isogram, isarithm, and isoline are generic terms embracing both isometric lines and isopleths. Isometric line refers to a line representing a constant value obtained from measurement at a series of points e.g. contour line. An isopleth is a line connecting points assumed to have equal values e.g. isodem.[2]

The following are selected from a much larger number of terms for special forms of the isogram. They are presented in chronological order according to their assumed first cartographic use. The contemporary terms, as listed, were not in every instance applied for their earliest cartographic usage. Each definition should be preceded by the words, "line along which..." and followed by "...is, or is assumed to be, the same or constant."[3]

Isobath	depth below a datum (e.g. mean sea level)
Isogonic line	magnetic declination
Isocline	magnetic dip (inclination) or angle of slope
Isohypse (contour)	elevation above a datum (e.g., mean sea level)
Isodynamic line	value of intensity or a component of the intensity of the magnetic field
Isotherm	temperature (usually average)
Isobar	atmospheric pressure (usually average)
Isohyet	precipitation
Isobront	occurrence of thunderstorms
Isanther	time of flowering of plants
Isopag	duration of ice cover
Isodem	population
Isoamplitude	amplitude of variation (often of annual temperature)
Isoseismal line	number (or intensity) of earthquake tremors

[1] Isogram, proposed in 1889 by Francis Galton, is the most general of these terms. It is defined as a line along which values are, or are assumed to be, constant.

[2] John K. Wright, "The Terminology of Certain Map Symbols," *Geographical Review,* vol. 34 (1944), 653–54.

[3] Werner Horn, "Die Geschichte der Isarithmenkarten," *Petermanns Geographische Mitteilungen,* vol. 103 (1959) 225–32; J. L. M. Gulley and K. A. Sinnhuber, "Isokartographie: Eine Terminological Studie," *Kartographische Nachrichten,* vol. 4 (1961) 89–99; The Royal Society, "Glossary of Technical Terms in Cartography" (London: The Royal Society, 1966).

Isochasm	annual frequency of aurorae
Isophot	intensity of light on a surface
Isoneph	degree of cloudiness (often average)
Isochrone	travel time from a given point
Isophene	date of beginning of a plant species entering a certain phenological phase
Isopectic	time of ice formation
Isotac	time of thawing
Isobase	vertical earth movement
Isohemeric line	minimum time of (freight) transportation
Isohel	average duration of sunshine in a specified time
Isodopane	cost of travel time
Isotim	the price for certain goods
Isoanabase	landrising
Isophort	freight rates on land
Isonau	freight rates at sea
Isomist	wages
Isothym	intensity of evaporation
Isoceph	cranial indices
Isochalaz	frequency of hail storms
Isogene	density of a genus
Isospecie	density of a species
Isodyn	economic attraction
Isohydrodynam	potential water power
Isostalak	intensity of plankton precipitation
Isovapor	vapor content in the air
Isodynam	traffic tension
Isohygrom	number of arid or humid months per year
Isobenth	amounts of zoo-benthos per unit area at given depths
Isonoet	equal average degree of intelligence

Appendix C

Glossary

The list of contemporary cartographic terms which follows was compiled from the text. Foreign words and proper names were excluded, as well as any except the most general projection and isoline terms which are the subject of Appendixes A and B, respectively. The list was then submitted to a group of geography students specializing in cartography. If a majority thought a term was already too well known to require definition, it was immediately eliminated from the list. The remaining terms were then compared with their usage in the text to see if they were adequately defined there; if so, they were eliminated. The meanings of those still remaining were looked up in a standard college dictionary, such as any student might possess, and if they were suitably defined for the purpose of this work they were dropped from the list. Hence, the Glossary contains only a selection of the technical terms used in this book. A number of sources were consulted in its compilation. However, many are adapted and a few taken directly from, *Glossary of Mapping, Charting, and Geodetic Terms,* (Washington, D.C.: Department of Defense, Department of the Army, Corps of Engineers, U.S. Army Topographic Command, 2nd edition, 1969); and a few adapted from *Glossary of Technical Terms in Cartography* (London: The Royal Society, 1966). These compilations contain a very large number of cartographic definitions and should be consulted for additional terms not included in the following list.

aerial survey, Mapping utilizing photographic, electronic, or other data obtained from an airborne station. Also called *air survey.*

altazimuthal theodolite, An instrument equipped with both horizontal and vertical graduated circles, for the simultaneous observation of horizontal and vertical directions or angles.

altitude tinting, See *hypsometric tinting.*

anaglyph, A stereogram in which the two views are printed or projected superimposed in complementary colors, usually red and blue. By viewing through filter spectacles of corresponding colors, a stereoscopic image is formed.

arc of the meridian, A part of an astronomic or geodetic line of longitude.

area proportional to (APT) map, A cartogram in which the surface extent of features is relative to the amount of the map data (e.g. population) rather than the geographical extent of the base to which it is related.

area reference grid, A plane-rectangular coordinate system usually based on, and mathematically adjusted to, a map projection and with numbers and/or letters used to designate positions of reference to the system.

area[l] symbol, A continuous and distinctive shading, tone, or repetitive pattern used on a map to represent features, real or theoretical, having usually considerable surface extent (e.g. forests, religion). It contrasts with a point symbol or line symbol (q.v.).

astronomical north, See *north.*

azimuth, The horizontal direction of a line measured clockwise from a reference plane, usually the meridian.

azimuthal map projection, A systematic representation of the graticule on which the directions of all lines radiating from a central point or pole, are the same as the directions of the corresponding lines on the graticule.

bar scale, See *graphical scale.*

base data, Fundamental cartographic information (e.g. coastlines, political boundaries) in relation to which additional data of a more specialized nature may be compiled or overprinted.

base line, A surveyed line established with more than usual care, to which surveys are referred for coordination and correlation.

block diagram, A representation of a landscape usually in perspective or isometric projection, frequently exaggerated in the vertical scale.

cadastral map, A plan showing the boundaries of subdivisions of land, usually with bearings and lengths and the areas of individual tracts, for purposes of describing and recording ownership.

cardinal direction, Any of the four principal astronomical directions on the surface of the earth: north, east, south, west.

cartobibliography, A systematic list of maps usually relating to a given region, subject, or person.

cartogram, An abstracted or simplified map for displaying quantitative data for which the base is normally not true to scale.

cartouche, A feature of a map or chart, often a decorative inset, containing the title, legend, or scale, or all of these items.

chorographic-scale map, A systematic representation of an intermediate-sized land area (e.g. a country) in contrast to a small-scale map or a large-scale map (q.v.). Also called an *intermediate-scale map.*

choropleth map, A systematic representation in which color or shading is applied to areas bounded by statistical or administrative limits.

color separation, The process of preparing a separate drawing, engraving, or negative for each color required in the production of a lithographed map or chart.

compass north, See *north.*

compass rose, A circle graduated from the reference direction, usually north, in compass points, or degrees (0–360) or both.

condensed projection, A systematic representation of the graticule with areas of little or no importance for a particular purpose eliminated and the remainder brought close together.

conformal map projection, A systematic representation of the graticule on which the shape of any small area of the surface mapped is unchanged; also called an *orthomorphic map projection,* it contrasts with an equal-area map projection (q.v.).

contour, An imaginary line connecting all points which are at the same elevation above or below a datum surface, usually mean sea level.

contour interval, The vertical distance between two adjacent contour lines.

controlled mosaic, An assemblage, usually of rectified aerial photographs, oriented and scaled to horizontal ground control.

coordinate system, A graticule, or a Cartesian grid, in which points are located from two (or three) axes which intersect at a point.

dasymetric map, A representation in which color or shading is applied to areas which have homogeneity, within specified limits, and in which it is not necessary for the color or shading to be limited by statistical or administrative boundaries.

datum, Any numerical or geometrical value, surface, line, or point which may serve as a base or reference for other quantities.

declination, See *magnetic declination.*

deformation, See *map distortion.*

density symbol, Shading, or color, used to cartographically represent quantity; usually the greater the amount, the deeper the shading or color.

dimensional stability, The ability of material to maintain size caused by changes in moisture content and temperature.

distortion, See *map distortion.*

dot map, A systematic representation of earth phenomena in which dots (usually of uniform size) each represent a specific number of the distribution being mapped.

equal-area map projection, A systematic representation of the graticule on which the area of any enclosed figure on the map is equal to the area of the corresponding figure on a globe of the same scale; it contrasts with a conformal projection (q.v.). Also called *equivalent map projection.*

field survey, See *ground survey.*

flow line, A linear cartographic symbol in which the width varies in proportion to the quantity being mapped.

form line, A linear symbol resembling a contour, but often broken or dashed, representing only approximate elevation and used to show the shape of the terrain rather than actual height.

four-color process, See *process color.*
fractional scale, See *representative fraction.*

general map, A systematic representation of an area showing a variety of geographical phenomena (e.g. coastlines, political boundaries, transportation lines) used for planning, location, reference, etc.; it contrasts with a thematic map (q.v.).
generating globe, A model of the sphere used for the development of perspective map projections, or a theoretical sphere to which projections may be referred for comparative purposes. The radius of the generating globe bears the same relationship to the earth as is denoted by the representative fraction of the resulting map.
geo-cartography, Mapping of earth phenomena in contrast to the mapping of extraterrestrial and other bodies.
geographical north, See *north.*
geographical scale, See *small-scale.*
globe gore, A lune-shaped segment which can be fitted to the surface of a sphere with little distortion or deformation.
graduated circle, A disc-shaped symbol proportional in actual area, or appearance, to the amount of the phenomena being mapped relative to other similarly shaped symbols. Also called a *proportional circle.*
graphical scale, A graduated line by means of which distances on a map or chart may be measured in terms of ground distances; also known as a *bar scale* or *linear scale.*
graticule, A network of lines representing the earth's parallels of latitude and meridians of longitude.
great circle, A line on the earth's surface, the plane of which passes through the center of the globe. This shortest distance between two points on the sphere is also known as an *orthodrome.*
grid, A (Cartesian) reference system of two sets of parallel lines intersecting at right angles forming squares; also used loosely of the (earth) graticule (q.v.).
ground survey, Measurement and mapping in the field, as distinguished from aerial survey (q.v.).

hachure, A short line running in the direction of maximum slope to indicate in relation to other such lines by thickness and spacing, the relief of the land.
high latitude, A polar or sub-polar area of the earth.
hypsometric tinting, A method of showing relief on maps and charts by coloring, in different shades, those parts which lie between different levels.

inset map, A separate map, usually of a different scale than the main map, positioned within the borders of a larger map.
intermediate-scale map, See *chorographic-scale map.*
interrupted map projection, A systematic representation of a graticule in which the origin or central meridian is repeated in order to reduce peripheral distortion; also known as *recentered projection.*
inverted image, See *psuedoscopic image.*
isarithm, isogram, isoline, See *Appendix B.*
isometric diagram, A representation simulating the third dimension, in which the scale is correct along three axes.

large-scale map, A systematic representation of a small land area (with a representative fraction arbitrarily set at 1:75,000 or greater); also sometimes called *topographic scale.*
layer tinting, See *hypsometric tinting.*
legend, An explanation of, or key to, the cartographic symbols used on a map, diagram, or model.
leveling, The operation of measuring vertical distances, directly or indirectly, to determine elevations.
libration, A real or apparent oscillatory motion, particularly on the moon. This results in more than half of the moon's surface being revealed to an observer on the earth even though the same side of the moon is always toward the earth.
line[linear] symbol, A distinctive line used to represent features, real or theoretical, which have length but little or no width (e.g. roads, political boundaries).
linear scale, See *graphical scale.*
low latitude, A tropical or sub-tropical area of the earth.
loxodrome, See *rhumb line.*

magnetic declination, The angle between the magnetic and geographical meridians at any place expressed in degrees, east or west, to indicate the direction of magnetic north from true north.
magnetic north, See *north.*
magnetic variation, Used as a synonym for magnetic declination but, more specifically, to indicate changes in this relationship within certain time limits i.e. temporal variation.
map data, Specific cartographic information plotted in relation to base data (q.v.).
map distortion, Alteration in shape owing to the transformation of the sphere or spheroid, or part of such a figure through projection on a plane surface; also called *map deformation.*
map projection, Any systematic arrangement of the meridians and parallels (graticule) of the allside curving figure of a sphere or spheroid, or a part of such a figure, on a plane surface.
mean sea level (MSL), The average height of the surface of the sea for all stages of the tide.
metes and bounds survey, The description of the boundaries of tracts of land (e.g. properties) by giving the bearing and length of each successive line, often keyed to an ownership list.
mid-latitude, An area between the sub-tropical and sub-polar areas of the earth.
mosaic, See *controlled mosaic* and *uncontrolled mosaic.*

natural scale, See *representative fraction.*
normal case of a projection, The mathematically simplest aspect of a representation of the graticule (typically principal directions in the representation coincide with those on the graticule). See *oblique map projection* and *transverse map projection.*
north, The primary reference direction relative to the earth. Magnetic or compass north is the direction of the north-seeking end of a magnetic compass needle, not subject to local disturbance. True, astronomical, or geographical north is the northern direction of the meridian at the point

of observation. Grid north is the direction of the north-south lines on a map coincident with true north only at the meridian of origin.

oblate spheroid, An ellipsoid of rotation, the shorter axis of which is the axis of rotation. The earth is approximately an oblate spheroid.

oblique map projection, A systematic representation of the graticule with an axis inclined at an angle between 0 and 90 degrees, but neither of those two specific angles. See *normal case of a projection* and *transverse map projection.*

orientation, The act of establishing, or the state of being in, correct relationship in direction with reference to the points of the compass.

orthodrome, See *great circle.*

orthomorphic map projection, See *conformal map projection.*

parallax, The apparent displacement of the position of a body, with respect to a reference point or system, caused by a shift in the point of observation.

perspective diagram, A representation simulating the third dimension, with the appearance to the eye of objects correct in respect to their relative distance and position.

photogrammetry, The science or art of obtaining reliable measurements, and/or preparing maps and charts, from aerial photographs using stereoscopic equipment and methods.

photolithography, A method of printing in which the original subject is photographed and the consequent image is transferred to a (metal) plate for lithographic printing. Loosely used for the whole process of *lithography,* and vice versa.

pie graph, Circular symbol divided into sectors to indicate proportions of a total value. Also known as a *sectored circle.*

planimetric map, A systematic representation of land with only the horizontal positions of features shown. Contrasts with a topographic map (q.v.).

plastic scribing, See *scribing.*

plastic shading, See *shaded relief.*

point symbol, A distinctive device used to represent features, real or theoretical, usually having limited areal extent (e.g. settlements). However, such symbols are sometimes used in combination with other such symbols to show density, e.g. dot map (q.v.).

process color, A photo mechanical method of printing in which the separation of the colors of the original is accomplished mechanically and photographically. It includes, as a special case, four color process in which filters and screens are used to break images into four colors (red, yellow, blue, and black) which, when recombined at the printing stage will simulate essentially all colors in the original.

profile, A vertical section of the surface of the earth and/or the underlying strata, along any fixed line. It often involves vertical exaggeration (q.v.).

prolate spheroid, An ellipsoid of rotation, the longer axis of which is the axis of rotation.

proportional circle, See *graduated circle.*

pseudoscopic image, A three-dimensional impression which is the reverse of that actually existing as in photographs, shading, etc. of relief. Also called an *inverted image.*

range line, In the United States Public Land Survey a boundary of a township (q.v.), surveyed in a north-south direction.

recentered map projection, See *interrupted map projection.*

reconnaissance map, The cartographic product of a preliminary examination or survey of an area and therefore of a lower order of accuracy than later more rigorous surveys.

remote sensing, The detection and/or the recording of data about an object without having the sensor in direct physical contact with the object.

representative fraction (RF), The scale of a map or chart expressed as a fraction or ratio which relates unit distance on the map to distance measured in the same unit on the ground e.g. 1:1,000,000. Also called a *natural scale* or *fractional scale.*

rhumb line, A line on the surface of the earth making the same angle with all meridians. Also called a *loxodrome* or *line of constant compass bearing,* it spirals toward the poles in a constant, true direction.

scale, The ratio of a distance on a map, globe, model, photograph, etc., to its corresponding distance on the ground or another graphical representation.

scribing, The process of preparing a negative (or positive) which can be reproduced by contact exposure. Portions of a photographically opaque coating are removed from a transparent (usually plastic) base with specially designed tools.

section, In the United States Public Land Survey the unit of subdivision of a township. Normally a quadrangle of one mile square, there are 36 such units in a township (q.v.).

sectored circle, See *pie graph.*

shaded relief, The rendering of landforms by continuous graded tone to give the appearance of shadows thrown by a light source normally located above the northwest of the map.

small-scale map, A systematic representation of a large land area; also called a *geographical-scale map.*

spherical coordinates, A system of polar coordinates in which the origin is in the center of the sphere and the points all lie on the surface. Also loosely known as a spherical grid.

spheroid, Any figure differing slightly from a sphere; in geodesy one of several mathematical figures closely approaching the undisturbed mean sea level of the earth extending continuously through the continents (geoid) used as a surface of reference for geodetic surveys.

spot elevation, A point on a map or chart marked usually by a dot, with a numerical expression of elevation; also called *spot height.*

standard line, A parallel, meridian or other basic linear feature of a map projection along which the scale is as stated on the map or chart, and which is used as a control line in the computation of a map projection. Also called *standard meridian* or *standard parallel.*

statistical surface, A theoretical three-dimensional figure resulting from isopleth, choropleth, or other forms of quantitative mapping.

stereoscope, A binocular optical instrument to assist an observer to view photographs and diagrams to obtain a mental impression of a three-dimensional model.

strip map, A cartogram showing, in diagrammatic form, routes, etc. from one point to another along a more or less straight line.

symbol, A diagram, design, letter, character, or abbreviation placed on maps, charts etc. which by convention, usage or reference to a legend is understood to stand for, or represent, a specific characteristic or feature. They may be in the form of an areal, linear, or point or other symbol (q.v.).

synoptic chart, A systematic representation to indicate conditions prevailing, or predicted to prevail over a considerable area at a given time e.g. weather map.

thematic map, A systematic representation of an area normally featuring a single distribution as its map data (e.g. population) and for which the base data serve only to help locate the distribution being mapped. In its function it contrasts with a general map (q.v.).

topographic map, A systematic representation of a small part of the land surface showing physical features (e.g. relief, hydrography), and cultural features (e.g. roads, administrative boundaries). These large-scale maps present both vertical and horizontal features in measurable form.

topographic-scale map, See *large-scale map.*

township, In the United States Public Land Survey a quadrangle of approximately 6 miles on a side consisting of 36 sections.

township line, In the United States Public Land Survey a boundary of a township (q.v.) surveyed in an east-west direction. See also *range line.*

transverse map projection, A systematic representation of the graticule with its axis rotated 90 degrees (right angles) to that considered as the normal case of a map projection (q.v.) in any particular example. See also *oblique map projection.*

uncontrolled mosaic, An assemblage of unrectified prints, the detail of which has been matched from print to print without ground control or other orientation.

variation, See *magnetic variation.*

verbal scale, An expression of the relationship between specific units of measure on the map and distance on the ground (e.g. one inch equals one mile); a less general expression than the representative fraction (q.v.), in this case 1:63,360.

vertical exaggeration, The change in a model surface or profile created by proportionally raising the apparent height of all points above the base level while retaining the same base.

volumetric symbol, A cartographic device (e.g. simulated sphere) to give a quantitative impression of the third dimension.

zenithal map projection, See *azimuthal map projection.*

For other definitions see the text; consult the index for specific page references.

Illustration Source List

FIGS. 1.1, 1.2, and 1.3 from Sir Henry Lyons, "The Sailing Charts of the Marshall Islanders," *The Geographical Journal,* Vol. 72 (1923), facing 327, 327, and 326, respectively. Reprinted by permission of the Royal Geographical Society.

FIG. 2.1 from the Department of Special Collections, Research Library, UCLA: Prince Youssouf Kamal, *Monumenta Cartographica Africae et Aegypti,* Tome I (1926), Fascicule I, 6. FIGS. 2.2 and 2.4 from Ekhard Unger, "Ancient Babylonian Maps and Plans," *Antiquity,* Vol. 9 (1935), facing 315 and 312, respectively. Reprinted by permission of the publisher. FIG. 2.3 from Theophile J. Meek, "The Orientation of Babylonian Maps," *Antiquity,* Vol. 10 (1936), plate VIII, facing 225. Reprinted by permission of the publisher.

FIGS. 3.1, 3.3, and 3.4 from Joseph Needham and Wang Ling, *Science and Civilization in China,* Vol. 3 (Cambridge: Cambridge University Press, 1959), facing 548; Fig. 227, facing 549; and Fig. 231, facing 552, respectively.

FIGS. 4.6, 4.7, and 4.8 from the Department of Special Collections, Research Library, UCLA: Prince Youssouf Kamal, *Monumenta Cartographica Africae et Aegypti,* Tome III (1934), Fascicule IV, 864, 858, 867, respectively.

FIG. 5.1 from the New York Public Library. FIG. 5.3 from E.G. Ravenstein, *Behaim's Globe* (London: George Philip & Son, Ltd., 1908). FIGS. 5.6 and 5.7 from the Department of Special Collections, Research Library, UCLA. Reprinted by permission.

FIG. 6.2, reprinted by permission of the Royal Society. FIGS. 6.3 and 6.4 reprinted by permission of the Royal Geographical Society. FIG. 6.5 from *Geschiedenis der Kartografie van Nederland,* by Sybrandus Johannes, Fockema Andreae, and B. van t'Hoff, s'Gravenhage (The Hague: Martinus Nijhoff, 1947), plate 14. Reprinted by permission of the publisher. FIG. 6.6 from Charles J. Singer, ed., *A History of Technology,* Vol. 4 (Oxford: Clarendon Press, 1954–58), 606. Reprinted by permission of the publisher. FIGS. 6.7 and 6.8 from the Department of Special Collections, Research Library, UCLA. Reprinted by permission.

FIGS. 7.1 and 7.2 from R.A. Skelton, "The Early Map Printer and His Problems," *The Penrose Annual,* Vol. 57 (1964), 185 and 184, respectively. Reprinted by permission. FIGS. 7.5 and 7.6 from Arthur H. Robinson, "The 1837 Maps of Henry Drury Harness," *The Geographical Journal,* Vol. 121 (1955), facing 448 and 441, respectively. Reprinted by permission of the author and the Royal Geographical Society. FIG. 7.7 from E.W. Gilbert, "Pioneer Maps of Health and Disease in England," *The Geographical Journal,* Vol. 124 (1958), 174. Reprinted by permission of the author and the Royal Geographical Society. FIG. 7.8 from Norman J.W. Thrower, "Relationship of Discordancy in Cartography," *International Yearbook of Cartography,* Vol. 6 (1966), 21. Reprinted by permission of the publisher, C. Bertelsmann Verlag. FIG. 7.9 from the National Archives, Washington, D. C. FIG. 7.10 from Norman J.W. Thrower, *Original Survey and Land Subdivision: A Comparative Study of the Form and Effect of Contrasting Cadastral Surveys* (New York: Rand McNally & Co., for the Association of American Geographers, 1966), p. 40. Reprinted by permission of the publisher. FIG. 7.11 from Norman J.W. Thrower, "The County Atlas of the United States," *Surveying and Mapping,* Vol. 21 (1961), p. 336. Reprinted by permission of the American Congress on Surveying and Mapping, Washington, D.C. FIG. 7.12 from William M. Davis, assisted by William H. Snyder, *Physical Geography* (New York: Ginn and Company, 1898), p. 170. FIG. 7.13 from the Department of Special Collections, Research Library, UCLA. Reprinted by permission.

FIG. 8.2 from Richard A. Gardiner, "A Re-Appraisal of the International Map of the World (IMW) on the Millionth Scale," *International Yearbook of Cartography,* Vol. 1 (1961), between 32–33. Reprinted by permission of the publisher, C. Bertelsmann Verlag. FIG. 8.7 from L. Dudley Stamp and E.C. Willatts, "The Land Utilization Survey of Britain: An Outline Description of the First Twelve One-Inch Maps," (1935), London School of Economics, frontispiece. Reprinted by permission of the publisher, Geographical Publication, Berkhamsted, England. FIG. 8.8 from the *Daily Weather Maps,* ESSA/EDA Weekly Series, December 29, 1969–January 4, 1970. FIG. 8.9 from the sheet "Mare Nectaris—Mare Imbrium" by the U.S. Army Map Service.

FIG. 9.1 from Norman J.W. Thrower, "California Population Distribution in 1960," Map Supplement No. 7, *Annals* of the Association of American Geographers, Vol. 56, No. 2 (June 1966). Reprinted by permission of the publisher. FIG. 9.4 from Arthur H. Robinson and Norman J.W. Thrower, "A New Method of Terrain Representation," *The Geographical Review,* Vol. 47 (1957), 513. Reprinted by permission of the publisher. FIG. 9.5 from Norman J.W. Thrower, "Extended Uses of the Method of Orthagonal Mapping of Traces of Parallel, Inclined Planes with a Surface, especially, Terrain," *International Yearbook of Cartography,* Vol. III (1963), Fig. 4, between 32–33. Reprinted by permission of the publisher, C. Bertelsmann Verlag. FIG. 9.6 from Hermann Bollmann, "New York Picture Map," Pictorial Maps, Inc., 97 Warren Street, New York City. Reprinted by permission of the publisher. FIG. 9.7 adapted from Lászlo Lackó, "The Form and Contents of Economic Maps," *Tijdschrift Voor Econ. En Soc. Geografie,* Vol. 58 (1967), 327–28. FIG. 9.9 from the University of California, Los Angeles, Campus Computing Network (UCLA–CCN).

Index

NOTE: Page references to illustrations, appendices, and the glossary are in *italics.*